THE
OBJECTIVIST'S
GUIDE
TO THE
GALAXY

THE OBJECTIVIST'S GUIDE TO THE GALAXY

ANSWERS TO THE ULTIMATE QUESTIONS OF LIFE, THE UNIVERSE, AND EVERYTHING

ZOLTAN CENDES, PhD

HOUNDSTOOTH
PRESS

THE OBJECTIVIST'S GUIDE TO THE GALAXY
Answers to the Ultimate Questions of Life, the Universe, and Everything
First Edition

ISBN   978-1-5445-4823-4   *Hardcover*
      978-1-5445-4822-7   *Paperback*
      978-1-5445-4824-1   *Ebook*
      978-1-5445-4825-8   *Audiobook*

For Marie
Let's explore the galaxy together.

CONTENTS

ABBREVIATIONS

AS—Ayn Rand. *Atlas Shrugged*. Random House, 1957.

FNI—Ayn Rand. *For the New Intellectual*. Random House, 1961.

HWK—Harry Binswanger. *How We Know: Epistemology on an Objectivist Foundation*. Tof, 2014.

ITOE—Ayn Rand. *Introduction to Objectivist Epistemology*. Penguin, 1966.

OPAR—Leonard Peikoff. *Objectivism: The Philosophy of Ayn Rand*. Plume, 1991.

PWNI—Ayn Rand. *Philosophy—Who Needs It?* Bobbs-Merrill, 1982.

RM—Ayn Rand. *The Romantic Manifesto*. World Publishing Company, 1969.

VOS—Ayn Rand. *Virtue of Selfishness*. New American Library, 1964.

INTRODUCTION

Forty-two. That's the answer to "The Ultimate Question of Life, the Universe, and Everything" provided in the book *The Hitchhiker's Guide to the Galaxy*, published by Douglas Adams in 1979.[1] *Forty-two.* As if the Universe could be reduced to a two-digit number.

While absurd, *The Hitchhiker's Guide to the Galaxy* has sold over 14 million copies, placing it in the top one hundred on the list of all-time bestselling books.[2] Adams meant "42" as a joke, apparently saying that "42 is the funniest of the two-digit numbers."[3] Yet the question is not absurd—we all wonder about "The Ultimate Questions of Life, the Universe, and Everything." This book intends to complete Adams's task: *The Objectivist's Guide to the Galaxy* provides scientific answers to life's ultimate questions about the Universe and ourselves.

1   Douglas Adams, *The Hitchhiker's Guide to the Galaxy* (Pocket Books, 1979).

2   Wikipedia, "List of Best-Selling Books," modified November 26, 2024, 16:10 (UTC), accessed December 3, 2024, https://en.wikipedia.org/wiki/List_of_best-selling_books.

3   John Lloyd, speech, 30th Anniversary Hitchhiker's recording at Douglas Adams Memorial Lecture, March 12, 2008, Royal Geographical Society.

Following Adams's lead, this book will tackle 42 Ultimate Questions. Each of these questions are derived from the three meta–Ultimate Questions—the three big, all-encompassing questions we ask ourselves about existence and the Universe.

Ultimate Question #1 is: *Why is there something rather than nothing?* For eons, the answer given by mystics has been *God*, as if that solved the problem. Positing a god does not explain why there is something rather than nothing; it merely moves the question into the unknowable category. Science deals with facts, and God is all about feelings, not facts.

The problem with Ultimate Question #1 is it is self-contradictory. Only a being in existence can ask the question: *Why is there something rather than nothing?* If there *was* nothing, you would not exist, and if you did not exist, you could not ask the question.

Thus, the answer to Ultimate Question #1 is existence exists, and you have no choice about it. Everything you will ever do, everything you will ever think, is derived from existence. And nonexistence—nothing—makes no contribution.

> It is self-contradictory for an existing being
> to question why existence exists.

Ultimate Question #2 is: *What is consciousness for?* The obvious answer is to move, act, and think in existence. But this answer has been shunned by philosophers through the centuries. Instead of accepting the obvious, they have fabricated fabulous tales: consciousness should contemplate an imaginary world of ideal Forms, a separate world where ideas exist divorced from the material things that we experience in our daily lives (Plato);[4] consciousness should

---

4   "Form," in *The Great Ideas: A Syntopicon of Great Books of the Western World (Vol. II)*, ed. Mortimer J. Adler and William Gorman (Encyclopædia...

obey God's thoughts, a nonmaterial and brainless God, a God consisting of pure thought (Augustine);[5] consciousness should "deny knowledge in order to make room for faith" (Kant);[6] consciousness should play "word games" (Wittgenstein).[7] The obvious answer makes more sense, as we will explore more fully in the book.

Ultimate Question #3 is: *What exists and how do we know it?* This question is harder than the first two. The first two questions concern metaphysics, the branch of philosophy that examines the basic structure of reality. Metaphysics is simple: Things exist, and you have no choice about it. But the third question, *What exists and how do we know it?*, is complex. Answering this question requires *knowledge*, and knowledge is the province of epistemology, the branch of philosophy concerned with the nature, origin, justification, and scope of knowledge. Here, mistakes are possible. We spend our lives trying to understand reality and often wonder if our thoughts are true or false.

Answers to Ultimate Question #3 in its various forms fill the bulk of this book. Establishing existence as the primary fact, and consciousness as the means to understand and operate in existence, is straightforward. But what exists? And how do we know it? That takes work.

The first part of Ultimate Question #3, *What exists?*, is the province of science. Science is the study of existents and their properties. Science has made remarkable progress during the past four centuries: Isaac Newton showed that a gravitational field permeates all space; John Dalton, James Clerk Maxwell, and others showed that matter is composed of atoms. Charles Darwin showed that man is

---

...Britannica, 1952), 526–42.

5   Augustine held that God is not material. Augustine, *Confessions*, book 7.1.1–2.

6   Immanuel Kant, *Critique of Pure Reason*, trans. and eds. Paul Guyer and Allen W. Wood (Cambridge University Press, 1999).

7   Ludwig Wittgenstein, *Philosophical Investigations*, trans. G. E. M. Anscombe (Macmillan, 1953).

the product of evolution; Albert Einstein showed that space and time are coupled. We describe some of these key scientific developments throughout the course of this book.

Answering the second part of Ultimate Question #3, *How do we know it?*, requires a guide—someone who has found a way through philosophy's epistemological maze. Only one philosopher has answered Ultimate Questions #1 and #2 correctly. Only one philosopher holds that existence exists independently of consciousness, and consciousness is the act of perceiving that which exists. Her name is Ayn Rand.

Ayn Rand is best known as the author of the novels *The Fountainhead* and *Atlas Shrugged*. She wrote afterward:

> The motive and purpose of my writing is the projection of an ideal man. The portrayal of a moral ideal, as my ultimate literary goal, as an end in itself—to which any didactic, intellectual or philosophical values contained in a novel are only the means.[8]

To portray the ideal man, Rand asked herself what it means to be moral, and on a deeper level, what is the logical, reality-based foundation of morality. This led to such questions as: What is real, what is imagined, and how does one know the difference? In answering these questions, Rand developed an entirely new philosophy—a philosophy fundamentally different from the subjectivism and the idealism of earlier philosophers. This allowed Rand to portray her ideal men, such as the architect Howard Roark, the inventor John Galt, and the business executive Dagny Taggart. They were real people living on earth, and their thoughts and actions were harmonious with life on earth.

---

8   Originally published in *The Objectivist Newsletter* in 1963, this essay is anthologized in RM, 145.

Rand first presented her philosophy, which she called *objectivism*, in her novel *Atlas Shrugged*. It is introduced in a speech by the hero, John Galt. While this speech runs to sixty pages, and would require three hours to deliver, it is nevertheless a condensation of her ideas. She spent the next twenty-five years of her life elaborating on her ideas, writing beautiful, logical, and profound works of nonfiction.

Leonard Peikoff defines objectivism as: the philosophy of Ayn Rand.[9]

This definition is circular—it reveals nothing about what objectivism *means*. Asked to present the essence of her philosophy "standing on one foot," Ayn Rand replied:

- *Metaphysics*: Objective reality
- *Epistemology*: Reason
- *Ethics*: Self-interest
- *Politics*: Capitalism[10]

However, an essence is not a definition. According to Rand, a definition must provide the *essential* distinguishing characteristic(s) of a concept. Such a definition is:

Objectivism—A philosophical system wherein all truth is the product of the logical identification of the facts of experience.[11]

This definition distinguishes objectivism from *intrinsicism*, the view that ideas exist independently of human consciousness, and

9   "Objectivism: The Philosophy of Ayn Rand (1991)," Ayn Rand Institute, accessed December 3, 2024, https://aynrand.org/novels/objectivism-the-philosophy-of-ayn-rand.
10   Ayn Rand, *The Objectivist Newsletter*, August 1962, 35.
11   See ITOE, 112.

from *subjectivism*, the view that ideas are created by our minds out of thought itself.[12] Objectivism is based on the *logical identification of things that exist in our experience.*

How do we identify things that exist? According to Rand:

> Whenever in doubt...about the standing of any concept, you can do what I have done in this discussion right now. I asked you, "What, in reality, does that concept refer to?"[13]

That's what we do. The search for Ultimate Answers begins by asking the question, "What in reality gives rise to that concept?" This requires science—the systematic study of the structure and behavior

---

12 In Ayn Rand's use, *intrinsicism* takes two forms: *realism* and *idealism*. Realists (Aristotle and his followers, most notably Thomas Aquinas) claim that ideas are embedded in the objects we see. Idealists (Plato, Augustine, and religious scholars in general) argue that ideas exist in an alternate universe, separate from the ordinary universe of everyday experience. In previous philosophic discourse, an intrinsic attribute is in the nature of the object (its mass, for example) while an extrinsic attribute is the way the object interacts with the universe (its weight, in this example). In nonobjective ethics, an intrinsic value is a value something has "in itself." Plato and others have argued that pleasure is intrinsically good, while pain is intrinsically bad (Plato, *Protagoras*, 353e.) Ayn Rand identified intrinsicism in epistemology as "'universals" inherent in things, versus subjectivism, "as products of man's consciousness, unrelated to the facts of reality," versus objective knowledge, where "concepts [are] ... the products of a cognitive method of classification whose processes must be performed by man, but whose content is dictated by reality." ITOE, 52–53; OPAR, 144; Dan Marshall and Brian Weatherson, "Intrinsic vs. Extrinsic Properties," *The Stanford Encyclopedia of Philosophy*, fall 2023 ed., eds. Edward N. Zalta and Uri Nodelman, https://plato.stanford.edu/entries/intrinsic-extrinsic/; and Michael J. Zimmerman and Ben Bradley, "Intrinsic vs. Extrinsic Value," *The Stanford Encyclopedia of Philosophy*, spring 2019 ed., ed. Edward N. Zalta, https://plato.stanford.edu/entries/value-intrinsic-extrinsic.

13 ITOE, 305.

of the natural world through observation and experiment.[14] But it requires more. It requires understanding the relationship between man's mind and the world. In short, it requires a philosophy that is compatible with "the facts of experience": objectivism.

According to objectivism, existence exists independently of consciousness; consciousness is awareness of existence. The "primacy of existence" route to knowledge requires that we place existence in the driver's seat and derive all facts through their correspondence with existents. The alternative, the "primacy of consciousness" approach to knowledge, is corrupt. Thoughts cannot be true if they do not correspond to things that exist in reality.

Based on this, we have derived the answers to the 42 Ultimate Questions using a combination of science and philosophy. Here is a sample of the Ultimate Questions that headline each chapter in this book: *What is consciousness for? What are concepts? Is infinity a number? Do we live in a clockwork universe? Are atoms real? Are scientific laws certain?* In answering these questions, we follow the scientific method within the wider context of objectivist thought. The first step in the combined scientific/objectivist method is observation. We begin by asking ourselves: "What, in reality, gives rise to the observation?" This is followed by developing a hypothesis, which is an explanation of the observation that can be tested against reality. In the scientific method, the hypothesis is modified iteratively until the predictions from the hypothesis agree with the observed results. The scientific/objectivist method provides a logical bridge between an observation and its understanding.

I remember how disappointed I was after reading Douglas Adams's answer, the number 42, to "The Ultimate Question of Life, the Universe, and Everything." Yes, *The Hitchhiker's Guide to the Galaxy* is

---

14   Dictionary.com, "science," accessed August 10, 2024, https://www.dictionary.com/browse/science.

science fiction. But much of the discussion is couched in philosophic terms and I had expected more. In hindsight, I realize *The Hitchhiker's Guide to the Galaxy* is a metanarrative, a postmodern term that claims true answers to Ultimate Questions are impossible to find. This book provides the objectivist's answers.

WHY IS THERE SOMETHING RATHER THAN NOTHING?

THE PRIMACY OF EXISTENCE

This question was answered in the introduction but bears more scrutiny. We noted the contradiction inherent in the question, *Why is there something rather than nothing?* If there were nothing, you could not ask the question because you would not exist. This argument takes consciousness as the primary concept. Since consciousness exists and consciousness is something, the "why is there something" question is self-resolving. The complementary argument, taking existence as primary, is derived by noting the word "nothing" in the question has no referent. Existence and nonexistence are mutually exclusive concepts. The *Why is there something?* question presumes the possibility of "nonexistence" as a state, but "nonexistence" does not exist; i.e., it is not a state. There is *always* something—you cannot get away from it. Everything begins with existence—even logic, which is how we can ask the question in the first place. So you cannot question existence, because existence is the very foundation upon which we are able to think at all.

This point is illustrated by René Descartes's famous statement, "I think, therefore I am"[1] and by Rand's inversion, "I am, therefore I'll think."[2] The foundational premise in Descartes's philosophy is "I think." He uses this premise to justify existence. Rand's foundational premise is "I am," which means existence, and she uses existence to justify thinking.

Ayn Rand defined the *primacy of existence* and the *primacy of consciousness* worldviews as follows:

> The primacy of existence (of reality) is the axiom that existence exists, i.e., that the universe exists independent of consciousness (of *any* consciousness), that things are what they are, that they possess a specific nature, an *identity*...The rejection of these axioms represents a reversal: the primacy of consciousness—the notion that the universe has no independent existence, that it is the product of a consciousness (either human or divine or both).[3]

You must choose. You can think of your mind as an instrument for *perceiving* reality, where your thoughts are the result of neural activity in your brain. Or you can think of your mind as an instrument for *creating* reality, where your thoughts are mystical entities, mingling with God's and others' thoughts to shape the galaxy.

---

1   René Descartes, *A Discourse on Method: Meditations and Principles*, trans. John Veitch (Orion Publishing Group, 2004), 15.
2   AS, 1058.
3   PWNI, 24.

The contradiction inherent in the question, *Why is there something rather than nothing?*, is easier to see from the primacy of existence worldview. Things exist—you have no choice about it. Your only choice is to function within existence. For you, consciousness without existence is a contradiction in terms.

With the primacy of consciousness worldview, the contradiction is hidden. Religious leaders say the answer to *Why is there something rather than nothing?*, is that God created the universe. This is a nonanswer. It does not explain why there is a God rather than nothing.

"Nothing" is not a possible state with either worldview.

In either case, the question, *Why is there something rather than nothing?*, is self-contradictory. There is always something. In the primacy of existence worldview, the something is existence—an existence that exists independently of whether it is observed by consciousness or not. In the primacy of consciousness worldview, the something is consciousness. It predates the existence of the galaxy it creates.

THE ANSWER

Why is there something rather than nothing? This question is self-contradictory: A being in existence must employ nothing—nonexistence—as a possible state in asking the question. One has no choice. If there was nothing, the question could not be asked. Since existence exists, a person cannot leave it, except through death, in which case existence continues, but the person's consciousness ceases to exist.

WHAT IS CONSCIOUSNESS FOR?

CONSCIOUSNESS EVOLVED "FOR SURVIVAL"

Five hundred and forty million years ago, a new creature roamed the earth. Named *Clementechiton sonorensis* by paleontologists, this stem mollusk is the earliest animal in the fossil record to have developed eyes and an associated brain.[1] Fossils do not record when consciousness first appeared on earth, but we can surmise that *Clementechiton sonorensis* had some degree of environmental awareness. Ayn Rand described consciousness as "the

---

1   Mark A. S. McMenamin, *Dynamic Paleontology: Using Quantification and Other Tools to Decipher the History of Life* (Springer, 2016); G. J. H. McCall, "The Vendian (Ediacaran) in the Geological Record: Enigmas in Geology's Prelude to the Cambrian Explosion," *Earth-Science Reviews* 77, no. 1 (July 2006): 1–229, https://doi.org/10.1016/j.earscirev.2005.08.004.

faculty of awareness—the faculty of perceiving that which exists."[2] In this case, *Clementechiton sonorensis* was one of the first animals to have been conscious.

Consciousness provides dramatic evolutionary advantages. Prior to *Clementechiton sonorensis*, many animals sensed light using photoreceptors and had "nerve nets" to guide their actions. But *Clementechiton sonorensis* had the first complex image-forming eyes and a nervous system to interpret the resulting vision. Evolution works through the survival of the fittest, and *Clementechiton sonorensis* was more fit to survive than its nonconscious counterparts. The development of consciousness led to the Cambrian explosion when major animal phyla first appeared.[3] The diverse categories of animals we know today evolved over a remarkably short period of time, roughly thirteen to twenty-five million years of the earth's 4.55 billion years of existence.[4] The evolution of consciousness, approximately 538.8 million years ago, had a remarkable effect on life on earth.[5] Oxford zoologist Andrew Parker says,

---

2   The dictionary definition of consciousness is "the state of being aware of and responsive to one's surroundings." As we shall see later, consciousness is an axiomatic concept that can only be defined ostensively. FNI, 124; and Dictionary.com, "consciousness," accessed December 3, 2024, https://www.dictionary.com/browse/consciousness.

3   Mammals appeared only 225 million years ago. Simon Conway Morris, *The Crucible of Creation: The Burgess Shale and the Rise of Animals* (Oxford University Press, 1998). T. S. Kemp, *The Origin and Evolution of Mammals* (Oxford University Press, 2005), 3.

4   Graham Budd, "At the Origin of Animals: The Revolutionary Cambrian Fossil Record," *Current Genomics* 14, no. 6 (2013): 344–54, https://doi.org/10.2174/1389 2029113149990011.

5   Shannon Hsieh et al., "The Phanerozoic Aftermath of the Cambrian Information Revolution: Sensory and Cognitive Complexity in Marine Faunas," *Paleobiology* 48, no. 3 (January 2022): 397–419, https://doi.org/10.1017/pab.2021.46.

Precambrian creatures were unable to see, making it difficult to find friend or foe. With the evolution of the eye, the size, shape, color, and behavior of animals was suddenly revealed. Once the lights were "turned on," there was enormous pressure to evolve hard external parts as defenses and clasping limbs to grab prey. The animal kingdom exploded into life, and the country of the blind became a teeming mass of hunters and hunted, all scrambling for their place on the evolutionary tree.[6]

Consciousness is awareness, but what does this mean? Is the humanoid robot Atlas, created by Boston Dynamics, conscious? Atlas uses depth sensors to detect its surroundings and generate point clouds of its environment, a control system to perform complex dynamic interactions involving its whole body and its environment, and twenty-eight hydraulic joints to provide high-power movement.[7] Among many other tasks, Atlas was programmed to dance to Berry Gordy, Jr.'s, "Do You Love Me? (Now That I Can Dance)." The effect is uncanny.[8] You would swear the robot is alive. Atlas seems to feel the music and sways to the music's beat. This robot dances just like us, we think.

Of course, the effect is an illusion. Atlas is not "aware." It has no thoughts or emotions. It was programmed by humans to move in mechanical ways. Atlas is not conscious in the sense that a dog or a cat is conscious. Why is that? What is the essential difference between a robot, a dog, or a cat?

---

6   Andrew Parker, *In the Blink of an Eye: How Vision Sparked the Big Bang of Evolution* (Perseus Books, 2003).

7   "Atlas and Beyond: The World's Most Dynamic Robots," Boston Dynamics, accessed December 3, 2024, https://www.bostondynamics.com/atlas.

8   "Do You Love Me?," Boston Dynamics, December 29, 2020, YouTube, 2:54, https://www.youtube.com/watch?v=fn3KWM1kuAw.

Figure 1. The humanoid robot Atlas taking a leap. *Image: Boston Dynamics*

In their seminal paper "What Is Consciousness For?," Lee Pierson and Monroe Trout argue that consciousness is more than simply an awareness of existence.[9] Consciousness is pointless if an animal's actions are automatic. An automaton with a predetermined response to sensory input is not conscious. A conscious being must focus its mind on dangers and opportunities and act accordingly to stay alive.

Volition—the ability to focus the mind and the eyes in one direction or another—is the fundamental difference between high-level animals and simpler life forms, such as plants, bacteria, and amoeba.

9   Lee M. Pierson and Monroe Trout, "What Is Consciousness For?," *New Ideas in Psychology* 47 (December 2017), 62–71, https://doi.org/10.1016/j.newideapsych.2017.05.004.

Under the Pierson-Trout hypothesis, consciousness involves more than just perception: *Consciousness is volitional control over awareness.* Volitional control over awareness in turn leads to volitional movement.

Consciousness evolved because of the adaptive benefits of volitional movement. Pierson and Trout write:

> There is no adaptive benefit to being conscious unless it leads to volitional movement. With volition, the organism is better able to direct its attention, and ultimately its movements, to whatever is most important for its survival and reproduction.[10]

A deer on the edge of a meadow will pause to consider its next steps. It is drawn to the delicious grass in the meadow but must be careful. There could be a predator lurking behind a rock or tree. It must focus its attention selectively to sort out any danger. It smells the air for the scent of a wolf or a panther; it listens to rustling leaves for the sound of a predator's paws. It enters the meadow only after determining no predator hides in the shadows.

Volitional attention evolved because the natural world is too complex to be automated. The data in the woods is overwhelming. A deer's selective focus is the key to its survival. Without volitional control over perception, it would be defeated by the complexity of life. According to Pierson and Trout:

> When we refer to "consciousness," we mean the functional unit of volitional attention and conscious experience. Conscious organisms (and *only* conscious organisms), informed by conscious experience, can volitionally direct their attention to control their movements.[11]

---

10 Pierson and Trout, "What Is Consciousness For?"
11 Pierson and Trout, "What Is Consciousness For?"

The Pierson-Trout hypothesis distinguishes animal cognition from the cognitive behavior of simpler life forms. "All living cells are cognitive," writes James A. Shapiro:

> Cognition is a basic feature of life because all living organisms have to adapt their physiology and behaviour to novel circumstances. Biological cognition means that cells are able to perceive changing features of their internal and external environment and undertake responses directed to survival, growth, and reproduction of themselves or their clonal relatives.[12]

In a review paper, Pamela Lyon argues that bacteria and humans share most cognitive capabilities, including sense perception, discrimination, memory, learning, problem-solving, communication, motivation, anticipation, awareness, and intentionality.[13] Notably absent from the list of shared cognitive capabilities is volition.

While Lyon speaks of the cognitive capabilities of bacteria, philosopher Michael Marder goes further. Marder argues in *Plant-Thinking: A Philosophy of Vegetal Life* that plants are conscious.[14] Plants make complex decisions, like when to bloom, how to grow roots toward water, and how to stretch leaves toward the sun. Plants developed their "knowledge" through countless generations of plant evolution. Marder concludes, "If consciousness literally means being 'with knowledge,' then plants fit the bill perfectly."[15]

---

12   James A. Shapiro, "All Living Cells Are Cognitive," *Biochemical and Biophysical Research Communications* 564 (July 2021), 134–49, https://doi.org/10.1016/j.bbrc.2020.08.120.

13   Pamela Lyon, "The Cognitive Cell: Bacterial Behavior Reconsidered," *Frontiers in Microbiology* 6 (April 2015), https://doi.org/10.3389/fmicb.2015.00264.

14   Michael Marder, *Plant-Thinking: A Philosophy of Vegetal Life* (Columbia University Press, 2013).

15   Ephra Livni, "A Debate Over Plant Consciousness Is Forcing Us to Confront...

While both bacterial cells and plants sense and respond to changes in external or internal conditions, their response is chemical, not neurological.[16] Neither bacteria nor plants have volition. Like the robot Atlas, their response is preprogrammed, by man in the case of Atlas, by evolution in the bacterial and plants' cases, through countless generations of success or failure at life.

Animal consciousness is different: It involves volitional control over awareness. Animals choose which path to follow, which food to eat, and which mate to copulate with. Volition separates animal consciousness from bacterial "cognitive capabilities" and plant "consciousness."[17]

THE ANSWER

What is consciousness for? Consciousness is a biological adaptation that aids in the organism's survival. While all organisms possess the ability to sense changes in their environment and respond to these changes, the responses of simple biological organisms such as bacteria and plants are determined by chemical processes. Consciousness goes further. A conscious being is not only aware of its surroundings but can make choices about which course of action to take.

...the Limitations of the Human Mind," *Quartz*, June 3, 2018, https://qz.com/1294941/a-debate-over-plant-consciousness-is-forcing-us-to-confront-the-limitations-of-the-human-mind.

16   Gáspár Jékely, "The Chemical Brain Hypothesis for the Origin of Nervous Systems," *Philosophical Transactions of the Royal Society B* 376 (2021), https://doi.org/10.1098/rstb.2019.0761.

17   Livni, "A Debate Over Plant Consciousness."

QUESTION 3

WHAT ARE CONCEPTS?

THE TREE OF KNOWLEDGE MUTATION

Homo sapiens entered the Chauvet Cave in southeastern France some 32,000 years ago.[1] They scraped the walls clear of debris and concretions, leaving a smoother and noticeably lighter area upon which to paint—at least thirteen different species, including lions, bears, deer, and rhinoceroses, appear on Chauvet's walls.[2] Figure 1 shows a pride of lions. The artwork is mesmerizing in its quality and three-dimensional effect. Abstract markings—lines and dots—are found throughout the cave. Some researchers

1   Catherine Ferrier et al., "Heated Walls of the Cave Chauvet-Pont d'Arc (Ardèche, France): Characterization and Chronology," *Paleo* 25 (2014): 59–78, https://doi.org10.4000/paleo.3009.
2   Benjamin Sadier et al., "Further Constraints on the Chauvet Cave Artwork Elaboration," *Proceedings of the National Academy of Sciences* 109, no. 21 (May 2012): 8002–6, https://doi.org/10.1073/pnas.1118593109.

hypothesize these markings are symbols, a form of "proto-writing."[3] The entrance to Chauvet Cave was sealed by a collapsing cliff 29,000 years ago, leaving the cave untouched until 1994, when it was explored by three spelunkers, including Jean-Marie Chauvet for whom it was named.[4] The Chauvet Cave contains some of the oldest and best-preserved figurative cave paintings in the world.[5]

Figure 1. Lions painted in the Chauvet Cave.[6]

3   Bennett Bacon et al., "An Upper Palaeolithic Proto-writing System and Phenological Calendar," *Cambridge Archaeological Journal* 33, no. 3 (2023): 1–19, https://doi.org/10.1017/S0959774322000415.

4   Zach Zorich, "A Chauvet Primer," *Archaeology* 64, no. 2 (March–April 2011): 39, https://archive.archaeology.org/1103/features/werner_herzog_chauvet_cave_ primer.html.

5   Jean-Marie Chauvet et al., *Dawn of Art: The Chauvet Cave: The Oldest Known Paintings in the World* (H.N. Abrams, 1996); and "Decorated Cave of Pont d'Arc, Known as Grotte Chauvet–Pont d'Arc, Ardèche," UNESCO, accessed December 4, 2024, https://whc.unesco.org/en/list/1426.

6   The original is in Ardèche, France. This is a replica of the painting from the Brno Museum, Anthropos, Czechia.

Homo sapiens had trod the earth for roughly 300,000 years at the time of the Chauvet Cave paintings.[7] From approximately 330,000 to roughly 30,000 years ago, *sapiens'* appearance was similar to that of modern man, yet cognitive ability was radically reduced. Talking with a *Homo sapiens* from this era would have been impossible for a modern human. Then, somewhat earlier than 30,000 years ago, a profound transformation unfolded. Noteworthy artifacts, such as cave paintings and fertility sculptures, emerged, signaling a shift in cognitive abilities and the onset of concept formation. According to Yuval Noah Harari, around 30,000 years ago, a momentous biological mutation occurred, commonly referred to as the Tree of Knowledge Mutation or the Cognitive Revolution. This pivotal event elevated *Homo sapiens* from being just another animal to becoming the dominant species on the planet.[8]

Concepts are as different from perceptions as algebra is from arithmetic.

Perceptions are of particular things; concepts are representations of all things of a certain type. Both involve volitional attention. An animal will focus its attention on one meadow and then another, never integrating its experience into an abstract mental unit. A human being will notice the similarities and differences between meadows and other landscapes and form the abstraction "meadow," a concept that applies to all meadows he or she will ever see. He or she does this using a process of measurement omission.

---

7   Aurélien Mounier and Marta Lahr, "Deciphering African Late Middle Pleistocene Hominin Diversity and the Origin of Our Species," *Nature Communications* 10, no. 1 (September 2019): 3406, https://doi.org/10.1038/s41467-019-11213-w.

8   Yuval Noah Harari, *Sapiens: A Brief History of Humankind* (Signal, 2014).

AYN RAND'S THEORY OF CONCEPT FORMATION

Existence slaps a newborn baby in the face. Sound and light fill the room, warmth and pressure come from the mother's embrace, pain and pleasure arise from hunger and nursing. At first, a baby experiences only sensations. After two months, the muscles of the eye strengthen. Even then, though its eyes can produce an image, the retina, and the brain circuitry, are too immature to transmit a clear image. The nervous system must develop enough to gather and process the sensory data before perception is possible.[9] A "perception" is a group of sensations retained and integrated by the brain of a living organism, which gives it the ability to be aware, not of single stimuli, but of entities, of things.[10]

Ayn Rand wrote:

> The building block of man's knowledge is the concept of *"existent"*...it [the existent] is implicit in every percept (to perceive a thing is to perceive that it exists) and man grasps it *implicitly* on the perceptual level...
>
> The (implicit) concept "existent" undergoes three stages of development in man's mind. The first stage is a child's awareness of objects, of things—which represents the (implicit) concept *"entity."* The second and closely allied stage is the awareness of specific, particular things which he can recognize and distinguish from the rest of his perceptual field—which represent the (implicit) concept "identity."

---

9   Velma Dobson and Davida Y. Teller, "Visual Acuity in Human Infants: A Review and Comparison of Behavioral and Electrophysiological Studies," *Vision Research* 18, no. 11 (1978): 1469–83, https://doi.org./10.1016/0042-6989(78)90001-9.

10   VOS, 19.

The third stage consists of grasping relationships among these entities by grasping the similarities and differences of their identities. This requires the transformation of the (implicit) concept "entity" into the (implicit) concept "unit."...

A unit is an existent regarded as a separate member of a group of two or more similar members...

The ability to regard entities as units is man's distinctive method of cognition.[11]

An infant will notice many things in its environment during the first years of its life: mother, father, beds, chairs, lights, etc. These items are entities available at the perceptual level. As the child grows, he or she will notice differences between them—for example, that beds are long and flat and people lie on them, while chairs are shorter, have backs, and people sit on them. This is the process of differentiation. Within the category of "bed," he or she will differentiate between beds: this bed, that bed, my bed, Mommy's bed. When the child thinks of a specific bed as distinguished from other beds, he or she has formed a unit.

This is where algebra comes in. Ayn Rand explained that concepts are formed by considering an entity and ignoring its particular measurements. For example, she describes the process of a child forming the concept "length":

If a child considers a match, a pencil and a stick, he observes that length is the attribute they have in common, but their specific lengths differ. The *difference* is one of *measurement*. In order to form the concept "length," the child's mind retains the attribute and omits its particular measurements...

Bear firmly in mind that the term "measurements omitted" does not mean, in this context, that measurements are regarded

---

11   ITOE, 5–6.

as non-existent; it means that *measurements exist, but are not specified*. That measurements *must* exist is an essential part of the process. The principle is: the relevant measurements must exist in *some* quantity, but may exist in *any* quantity.[12]

To form the concept "bed," a child observes that beds differ from other entities in that they are raised above the ground, have flat surfaces, and people use them to lie on. He or she keeps these characteristics and omits all the measurements: the height above the ground, the size of the lying surface, the color, and softness of the fabric, etc. This is a process of integration where all the units of "bed" in his experience are blended together into a single new mental entity. Lastly, the child must label the concept with a word, in this case "bed." The function of the word is to reduce the enormous complexity of the concept formed into a single concrete mental unit. Without a word label, the child would have to repeatedly differentiate beds from other things, isolate their characteristics, ignore their measurements, and combine the results.

Ayn Rand's definition of "concept" is:

> A concept is a mental integration of two or more units possessing the same distinguishing characteristic(s), with their particular measurements omitted, and united by a specific definition.[13]

A concept provides a relationship between our minds and reality. We must form this concept through a voluntary act called thinking. Objective knowledge is obtained by focusing on existents and forming concepts consistent with things that exist.

---

12   ITOE, 11–12.
13   ITOE, 10, 13.

Prior to Ayn Rand, philosophers attempted to find concepts "out there" in existence, or "in here" in our brains. "Out there" philosophers hold either that ideas live in a nonmaterial world of "ideal forms" outside of the human brain (Plato, Augustine, and religious scholars) or are present as essences in the objects themselves (Aristotle, Thomas Aquinas). Ayn Rand dubbed "out there" philosophies "intrinsicism." Intrinsic philosophers hold that one can passively receive ideas directly from the world of ideal forms, or from God, or from the material world. "In here" philosophers hold our minds create ideas internally using our own resources (Hume, Kant, linguistic analysts, and most modern philosophers). Called *subjectivism*, these philosophers hold that "true" reality is unknowable.

The intrinsic/subjective philosophies of knowledge present a false dichotomy. Concept formation evolved as a mechanism to aid humans in dealing with reality. It requires both observation of reality and voluntary mental processes to form a thought. Ideas are neither intrinsic nor subjective; they are objective.

We say a thought is true if it corresponds with reality.

An idea that is in variance with reality is said to be false. False ideas come in two forms: mistakes and fiction. Mistakes happen because the thought-forming process is fallible. Fiction is a deliberate attempt by man to further his life by examining alternate possibilities for living. Whether an idea is true or false, the ultimate arbiter of the correctness of the content of a consciousness is its interaction with external stimuli.

THE ANSWER

What are concepts? Concepts are generalizations of perceptual data. The key element in concept formation is measurement omission. A child sees similarities and differences between entities and focuses on one aspect of existents as opposed to other aspects. For example, the concept length is formed by considering such objects as a match, a pencil, and a stick, and noticing that length is a common attribute of each but the specific lengths differ. To form the concept "length," the child keeps the attribute but omits the specific measurement of its length. A word must label the new concept if it is kept as an element of knowledge.

WHY IS LANGUAGE MEANINGFUL?

CONCEPTUAL INTELLIGENCE

n his book *Out of Our Minds: What We Think and How We Came to Think It*, Felipe Fernández-Armesto says the difference between people and animals is imagination:

The term that best denotes what is special about human thinking is probably "imagination"...the power of seeing what is not there...[1]

In *Sapiens: A Brief History of Humankind*, Yuval Noah Harari attributes the difference between people and animals to our ability "to speak about fictions":

---

1   Felipe Fernández-Armesto, *Out of Our Minds: What We Think and How We Came to Think It* (University of California Press, 2019), 10.

Legends, myths, gods and religions appeared for the first time with the Cognitive Revolution. Many animals and human species could previously say, "Careful! A lion!" Thanks to the Cognitive Revolution, Homo sapiens acquired the ability to say, "The lion is the guardian spirit of our tribe." This ability to speak about fictions is the most unique feature of Sapiens language.[2]

What gives man "the ability to speak about fictions"? The text does not offer an answer. Is "imagination" unique to humans? Not according to Fernández-Armesto. As Fernández-Armesto points out, imagination consists of memory, anticipation, and recall, and these aspects of consciousness are not unique to human beings.[3] Some animals are capable of complex cognitive tasks such as using tools and possess memories superior to ours: Chimpanzees use stones to crack open nuts, New Caledonian crows manufacture probes out of twigs to impale larvae, scrub jays recall the locations of hundreds of food caches they hid months earlier, rats retrace routes in complex labyrinths that would confound a human, and pigeons retain hundreds of visual patterns to home in on their own lofts after long absences. Anticipation is highly developed in the wild where predator and prey both need it to anticipate the movements of the other. Animals also use sounds as "words" throughout the animal kingdom, warning of approaching danger or attracting a mate. Certain animals,

---

2   Harari, *Sapiens*, 24.

3   The word "word" is an essential element in concept formation, and generally means the label humans attach to concepts as explained above. However, researchers in animal behavior apply "word" to calls animals make to communicate with other animals, such as warnings of danger or calls of attraction, and to sounds and signs men use to train animals. This is unfortunate because the use of the word "word" in animal communication implies that animals form concepts, which is not the case. Animals communicate on the perceptual level through sounds and other signals.

such as Nim Chimpsky,[4] a chimpanzee, and Alex,[5] an African gray parrot, have learned hundreds of words to associate with particular objects or needs.

However, Herbert Terrace, the primary individual teaching words in sign language to the chimpanzee Nim Chimpsky, subsequently concluded that the chimpanzee did not show any meaningful sequential behavior. Nim's use of language provided only perceptual-level identifications. Terrace concluded that word use by animals to obtain rewards was strictly pragmatic and did not imply concept formation in the animal.[6]

Anyone who has observed a dog on a leash outside a store waiting for its master to come out knows the dog remembers its master went into the store, anticipates his master's return, and will recognize his master's voice when it is called. The dog can clearly "imagine" its master's return; it has "the power of seeing what is not there."

Percepts are concretes: This man is this man, that tree is that tree. An animal can focus its attention on one particular man, or one particular tree, but the perception of the man or of the tree is given by nature. An animal has no choice about what it perceives. Concepts are error prone and fungible. The concept "man" can be applied to any man or even to a nonman. A cartoon may feature a walking, talking cartoon tree. A walking, talking tree is a metaphysical impossibility; it is a fiction that only man can create.

---

4   Herbert S. Terrace et al., "Can an Ape Create a Sentence?," *Science* 206, no. 4421 (November 1979): 891–902, https://doi.org/10.1126/science.504995.

5   Dinitia Smith, "A Thinking Bird or Just Another Birdbrain?," *New York Times*, October 9, 1999, https://www.nytimes.com/1999/10/09/arts/a-thinking-bird-or-just-another-birdbrain.html?showabstract=1; and Irene Pepperberg, "Talking with Alex: Logic and Speech in Parrots," *Scientific American* (1998): 60–65.

6   Herbert S. Terrace, *Why Chimpanzees Can't Learn Language and Only Humans Can* (Columbia University Press, 2019).

While imagination and fictions are key aspects of human intelligence, they are not the primary component of human thought.

Man is unique in thinking in terms of *concepts*.

Concepts differ from percepts because they involve the Tree of Knowledge Mutation, alternatively called the Cognitive Revolution (see *Question 3—What are concepts?*). This mutation allows people to classify entities and actions in terms of their distinguishing characteristics, understand that the measurements of these characteristics can take on any value, integrate the resulting mental entity into a unified whole, and assign a word to the result. Our ability "to speak about fictions" is a consequence of our ability to conceptualize.

An animal such as a dog thinks on the perceptual level. To a dog, each human being is a specific individual with specific attributes. The dog does not form the concept "man," a concept applicable to all men alive today, all men who ever lived, or who will be alive in the future. Animals have "perceptual-level" imaginings of specific things.

Concepts alter the way we process percepts. When a dog sees a saltshaker sitting on a table, its perception differs fundamentally from a human's perception. The dog sees an object, possibly to be sniffed and examined, probably to be ignored. You see a device for shaking salt out of holes in the top of the container to season your food. It is not possible for a normal, modern human to perceive a saltshaker as a dog does, or as a baby does, as just an object to be examined. Your perception of the saltshaker is tied directly to the concept "saltshaker" you developed growing up.

By testing subjects who were asked to pick up an arbitrary object on a table as the lights were turned on, researchers revealed that grown-ups cannot separate the "concept saltshaker" from the "animal or baby perceived saltshaker." It takes only 140 milliseconds after the appearance of the object to recognize it, and 250 milliseconds to recognize how to use it.[7] Upon seeing a saltshaker, you do not go through a series of conceptual thoughts—gee, it is a glass container with a metal top, the top has perforations in it, it looks like grains of salt inside the container, so it must be a saltshaker. No, your perception of the saltshaker and your concept of a saltshaker are identical. You grasp the saltshaker in 390 milliseconds *in a manner that lets you shake salt.* If a cup had appeared on the table, your grip would be different.

A similar process occurs when you read. Suppose I write the word "salt" in this sentence. When you think of the word "salt," your mind instantly creates the concept "a white crystalline substance utilized for flavoring food." When you learn to read, you sound out each of the letters, "s," "a," "l," "t" separately, you think about it, and then figure out what it means. After reading the word "salt" dozens of times, the process becomes automatic. It is impossible for an advanced reader to see the word "salt" without simultaneously thinking of its meaning. When an animal sees "salt" written on a sign, the lines form a meaningless squiggle on the board; for you, "salt" is a concept. Immaculate perception occurs only with animals and infants. Percepts are informed by concepts in healthy normal adults.

Because we think in terms of concepts, we can often identify existents correctly from partial information. Consider the following meme:

Yuo cna porbalby raed tihs esaliy desptie teh msispeillgns.[8]

---

7 Scott Grafton, *Physical Intelligence: The Science of How the Body and the Mind Guide Each Other Through Life* (Pantheon, 2020), 154.

8 Laura Moss, "Why Your Brain Can Read Jumbled Letters," Treehugger, August 23, 2024, https://www.treehugger.com/why-your-brain-can-read-jumbled-letters-4864305.

Our brains can identify the concept "You" for the first word even though it is misspelled. Similarly with the other words. We use context to make predictions about what's to come. In the above meme:

1. The first and last letters of most words are correct and in the correct places.
2. All the other letters in the word are between them, in random order.

This trick does not work if the reordered letters form another correct word. For example, it does not work with "salt" because our minds form a different concept when we read "slat."

Testing subjects with images, parts of which were obstructed with a white square, Dr. Lars Muckli, from the University of Glasgow's Institute of Psychology and Neuroscience, observed: "We are continuously anticipating what we will see, hear or feel next. If parts of an image are obstructed we still have precise expectation of what the whole object will look like."[9]

Concepts are what distinguish people from animals. Only man forms concepts; only man can form the concept of a walking, talking tree. And, unlike percepts that appear automatically as soon as an animal focuses on the subject, concept formation is not automatic. To form a concept, a man must focus his mind on the similarities and differences of an existent with other entities, apply the process of measurement omission, and label the concept with a word. This process takes work. To understand man, it is critical to understand that concept formation is a volitional process that requires focus and effort.

---

9   University of Glasgow, "What Our Eyes Can't See, the Brain Fills In," news release, April 4, 2011, https://medicalxpress.com/news/2011-04-eyes-brain.html.

Ayn Rand's theory of concept formation is largely unknown in the scientific community. For this reason, behavioral scientists claim that many animals, from apes to bees, think in terms of concepts. For example, bees were put in a maze with differing visual markings above and below a reference line as shown in Figure 1. If a bee chose the wrong arm, it received a quinine solution, a strongly aversive stimulus for bees; if a bee chose the correct arm, the bee received a sweet reward. The arm of the maze containing the reward changed unpredictably from trial to trial, and the only way bees could identify the correct arm was by choosing the correct arrangement of two visual stimuli on the back walls of the Y-maze. The bees swiftly learned this task and could even pass a transfer test when presented with a completely unfamiliar target.[10]

The authors concluded the "bees had indeed conceptualized the relationship of above-ness and below-ness." Further experiments with honeybees tested their ability to determine sameness/difference relationships in visual patterns. Martin Giurfa wrote: "The experiments reviewed cannot be accounted for by low-level strategies and challenge, therefore, the traditional view attributing supremacy to larger brains when it comes to the elaboration of concepts."[11]

The definition of "concept" used by behavioral scientists is:

A concept of x is a body of information about x that is stored in long-term memory and that is used by default in the processes

---

10 Lars Chittka and Keith Jensen, "Animal Cognition: Concepts from Apes to Bees," *Current Biology* 21, no. 3, (February 2011): R116–R119, https://doi.org/10.1016/j.cub.2010.12.045.

11 Martin Giurfa, "Learning of Sameness/Difference Relationships by Honey Bees: Performance, Strategies and Ecological Context," *Current Opinion in Behavioral Sciences* 37 (February 2021): 1–6, https://doi.org/10.1016/j.cobeha.2020.05.008.

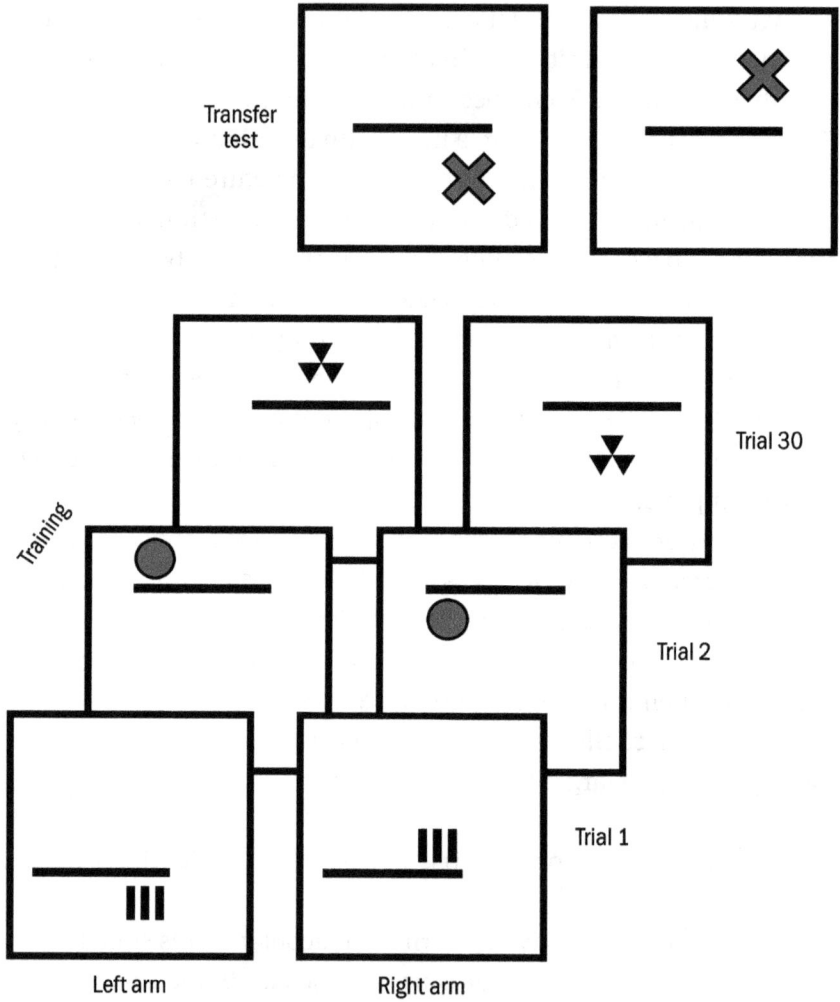

Figure 1. Stimuli sequentially shown to bees on the back wall of a Y-maze. Each pair of patterns simultaneously presented in the Y-maze contained the same visual stimuli, only distinguished by whether the "target" occurred below or above the line, which was always the same.[12]

---

12   Chittka, Jenson, "Animal Cognition."

underlying most, if not all, higher cognitive competences when they result in judgments about x.[13]

The key element here is, "used by default in...higher cognitive competences." Animals trained to perform specific tasks are thereby said to be using concepts. Researchers say that bees trained to navigate mazes with marks above or below a line have formed the concepts of above and below. If bees have concepts, we can talk to the bees!

All this would make sense if pattern recognition was a conceptual mental activity rather than perception based. Computer systems use artificial neural networks (ANNs) and deep learning, inspired by biological systems, to perform image analysis.[14] ANNs can easily sort images of shapes above a line from images of shapes below a line but no one would claim that the ANN has formed the concepts "above" and "below." Concepts are more than "higher cognitive competences"; they are abstractions formed through measurement omission and labeled by a word.

In 1904, Wilhelm von Osten asked his horse, Clever Hans, "If the eighth day of the month comes on a Tuesday, what is the date of the following Friday?" Hans answered by tapping his hoof eleven times. Clever Hans became world famous because he could add, subtract, multiply, divide, keep track of the calendar, and understand German. Osten never charged admission to demonstrations of Hans's intellectual abilities.[15]

---

13   Edouard Machery, "Précis of *Doing Without Concepts*," *Behavioral and Brain Sciences* 33, no. 2–3 (June 2010): 195–206, https://doi.org/10.1017/S0140525X09991531.

14   We discuss this further in *Question 31—How does my brain work?*

15   Edward T. Heyn, "Berlin's Wonderful Horse: He Can Do Almost Everything but Talk—How He Was Taught," *New York Times*, September 4, 1904, archived at https://timesmachine.nytimes.com/timesmachine/1904/09/04/101396572.pdf.

A commission, appointed by the German board of education, found that no tricks were involved in Hans's performance. Hans would get the correct answer even if von Osten himself did not ask the questions, ruling out the possibility of fraud. Finally, in 1907, biologist Oskar Pfungst found the horse gave the correct answer only when the questioner knew what the answer was, and the horse could see the questioner. He observed when von Osten knew the answers to the questions, Hans got 89 percent of the answers correct, but when von Osten did not know the answers to the questions, Hans answered only 6 percent of the questions correctly. Pfungst then examined the behavior of the questioner and showed as the horse's taps approached the right answer, the questioner's posture and facial expression indicated an increase in tension, which was released when the horse made the final, correct tap. This provided the cue the horse used to stop tapping.

Concepts are as different from percepts as horses are from bees. Bees can be trained to target visual patterns, and horses can learn the emotional states of their riders, but neither form concepts. Man is unique in speaking in terms of abstractions. Even *Homo sapiens* lacked the ability to form concepts until approximately 30,000 years ago. Man thinks in terms of concepts while animals are limited to the perceptual level of awareness.

THE ANSWER

Why is language meaningful? Language is meaningful because people speak in terms of concepts. Concepts are formed by focusing the human brain on existents, differentiating a class of existents from all others through their common attributes, omitting the measurements of these common attributes, integrating the result into a unified whole, and labeling the new concept with a word. Each word corresponds to something that exists. Stringing words together logically provides new information about existents.

WHAT IS THE RELATIONSHIP BETWEEN MY MIND AND THE WORLD?

OUR MULTIFACETED BRAINS

Man's mind is more than thoughts. Our minds regulate functions like heart rate or blushing without our conscious awareness, while other actions like breathing and walking are automated but can be consciously overridden. Our multifaceted brains have three different relationships with the world: physical, perceptual, and conceptual.

PHYSICAL INTELLIGENCE

You wake up feeling chilled in the middle of the night. You stand up in the dark, walk a few steps across the room, and turn the thermostat

THE OBJECTIVIST'S GUIDE TO THE GALAXY

up. Now you need to go to the bathroom, so you do that. Afterward you notice you are thirsty and hungry, so you go turn on the light and head down the hall to the refrigerator in the kitchen.

Which senses did you use?

Not hearing, smell, taste, or sight, at least not before you turned the light on. Touch is ever present when you lie down or sit, and when your feet step on the ground. But there is so much more. None of the sensations—feeling chilled, the need to use the bathroom, balance and walking in the dark, being thirsty and hungry—is provided by the five traditional senses. You need balance but clues to balance are not available from your eyes in the dark. To walk, you need to know where your foot is as you raise it, move it forward, and place it on the ground. Feeling chilled, bladder pressure, thirst, and hunger are relayed from sensors in various locations in your body to your brain stem, where a host of reflexes adjust blood flow to the brain, blood pressure, heart rate, and respiration. Getting up out of the bed is a conscious decision, but much of the rest is automatic.

Your sense of balance absent light is located in the vestibular system, a region of the inner ear where three semicircular canals converge. Your sense of limb velocity and movement, limb load, and limb limits are located in the proprioception system. Proprioceptors are mechanosensory neurons located within muscles, tendons, and joints. Touch is sensory data of pressure, texture, and temperature produced by external stimuli from the environment affecting your body. The proprioception system provides sensory data internal to the body itself.

The human body is wired with sensors. Of the 350,000 axons (the individual "wires" that carry information for each neuron) passing into the arm, 315,000 are sensory neurons, allowing a person to feel position, touch, temperature, and pain. Only 35,000 carry motor signals from the spinal cord to the arm muscles. A baby must integrate all these sensations into an integrated whole to form a perception. In his book, *Physical Intelligence: The Science of How the Body and the Mind Guide Each Other Through Life*, Scott Grafton writes:

> The vigilant mind requires something very fundamental of the brain: it has to create a concept of space to be able to both understand and act in the physical world... Physical intelligence...is foundational, a kind of knowing that frames much of what the mind spends its time engaged in.[1]

J. J. Gibson coined the term *affordances* to describe the way our brains engage with space. Consider a fast hike down a mountain trail. You focus on the path ahead, looking out for rocks and roots, willing your eyes to look from rock to rock, not letting them wander to the scenery, lest you trip on an obstacle. Your feet and arms move automatically, following the unconscious commands issued by your brain, but there is nothing unconscious about where you look. Perceptual awareness can only automate that which is expected: Your foot moves automatically to step over the rock you saw a second ago. But no "perceptual-level algorithm" exists to predict what will appear next along the trail. By the nature of reality, your eyes cannot be on autopilot. A lapse in your attention can cause you to stumble and fall.

Functional MRI (fMRI) systems have made the scientific study of brain activity possible. Scientists assume increased blood flow to a brain area indicates increased neural activity in that region. The resolution of this procedure is on the order of millimeters. A typical voxel (the three-dimensional analog of a pixel) contains a few million neurons and tens of billions of synapses.

fMRI experiments have revealed that visual attention is located in the upper regions of the brain called the *dorsal attention network*.

---

1   Grafton, *Physical Intelligence*, 6.

This network is closely tied to regions of the frontal lobe where neurons that control eye muscle movement are located. A separate syndicate of brain cells called the *ventral attention network*, located in the lower part of the brain, filters the data that gets through. Controlling the balance between the dorsal attention network and the ventral attention network is the "salience network."[2] The salience network shifts the focus of the brain's attention from one network to another and integrates attention with goals.

Drs. Wilder Penfield and Edwin Boldrey, of McGill University in Montreal, Canada, developed a distorted representation of the human body based on a neurological map of the areas and proportions of the human brain.[3] Called a *cortical homunculus* (from the Latin homunculus, "little man"), their drawings provide a conceptual map illustrating the brain regions dedicated to processing motor functions and sensory functions for different parts of the body. In Figure 1, the real estate on the cerebral cortex devoted to control of the hands is much greater than that dedicated to the feet.

The brain distributes the regulation of salience and attention throughout its various regions. Stroke victims may lose functionality, such as spatial reasoning and verbal command, but still maintain salience. Even brain damage does not eliminate the ability to focus.

On the perceptual level, our consciousness directs our actions; on the conceptual level, our consciousness directs our consciousness.[4]

2   Salience is defined as the property of being particularly noticeable or important.
3   Wilder Penfield and Edwin Boldrey, "Somatic Motor and Sensory Representation in the Cerebral Cortex of Man as Studied by Electrical Stimulation," *Brain* 60, no. 4 (December 1937): 389–443, https://doi.org/10.1093/brain/60.4.389.
4   Betsy Speicher, *The WHYS Way to Success and Happiness* (Sherwood Oaks Press, 2015).

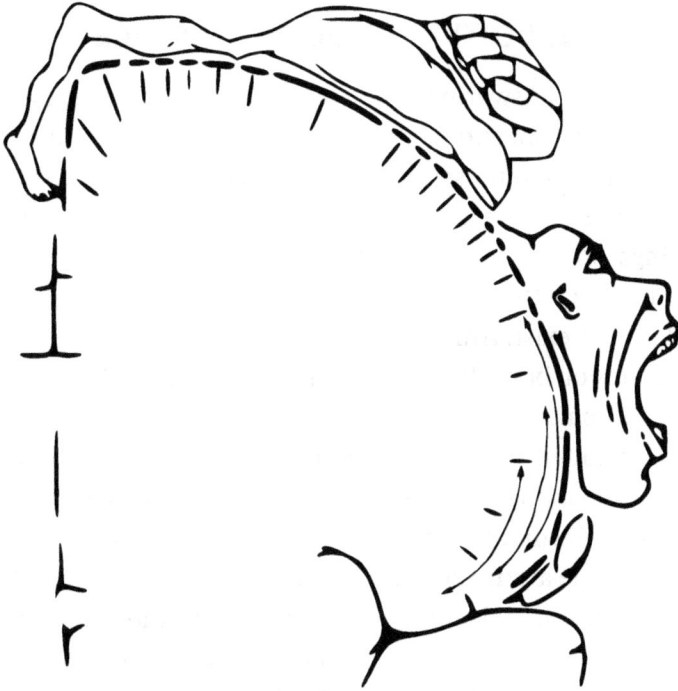

Figure 1. A cortical homunculus provides a distorted representation
of the human body based on a neurological "map" of the areas
and proportions of the human brain dedicated to processing motor
functions and sensory functions for different parts of the body.[5]

Man operates on both levels. Your decision to hike down a mountain trail is conceptual, but the actual process of hiking is perceptual. You do not think conceptually when running down a mountain trail. You choose where to look, whether to move to the left or right, speed up, slow down, or stop. Your feet move automatically, but there is nothing automatic about where you focus your attention. You are the agent in control of your life.

5   W. Penfield and T. Rasmussen, The Cerebral Cortex of Man: A Study of
     Localization Function (Macmillan, 1950), 199, fig. 26b.

PERCEPTUAL LEVEL INTELLIGENCE

As discussed in answering *Question 3—What are concepts?*, man's ability to form concepts is unique to humans. Concepts are algebra-like thoughts, representing unlimited entities of a certain type, while percepts are the arithmetic of thought representing only specific things. Conceptual intelligence is far superior to perceptual intelligence, but this does not mean that perceptual-level intelligence does not exist. Animals think only in terms of percepts, while man operates on both the perceptual and conceptual levels.

Both animals and humans require knowledge to stay alive: "Knowledge, for any conscious organism, is the means of survival."[6]

Knowledge comes in two forms: innate knowledge and learned knowledge. All animals possess innate knowledge, some to a remarkable degree. The arctic tern migrates from the Arctic and sub-Arctic regions of Europe, Asia, and North America to the Antarctic coast, and back again, traveling approximately 90,000 kilometers (56,000 miles) every year.[7] Monarch butterflies migrate from Central Mexico to the Western United States, to the Eastern United States, Southern Canada, and back to Mexico and, with a typical lifespan of a few weeks, accomplish this migration over four or five generations every year.[8] Salmon hatch in freshwater streams, stay for six months to three years, migrate to the oceans, stay in the oceans for one to five years, and return primarily to the streams where they were hatched to spawn the next generation.[9] The amount of coded

6 VOS, 21.
7 "Arctic Tern," All About Birds, Cornell Lab of Ornithology, accessed December 4, 2024, https://www.allaboutbirds.org/guide/Arctic_Tern/id.
8 Anurag Agrawal, *Monarchs and Milkweed: A Migrating Butterfly, a Poisonous Plant, and Their Remarkable Story of Coevolution* (Princeton University Press, 2017).
9 Hiroshi Ueda, "Physiological Mechanism of Homing Migration in Pacific...

information in the animal's genes required to accomplish such travels boggles the mind![10]

Higher-level animals need to learn skills to survive. Anyone who has owned a dog knows dogs can learn behaviors and tricks. A dog will learn at a young age that food and shelter are found at home. Assuming it is free to roam, the dog will sniff around its neighborhood, and learn where different things are, and perhaps chase a rabbit, but it knows where home is. When it is tired, hungry, or knows its master is coming home, it will head for home. Many a child has felt the joy of coming home to a wagging, welcoming dog, a dog expecting the child arriving home at the same time each school day.

Mammals are raised by their mothers. Young mammals learn survival skills from their mothers, from life experience, and sometimes with the aid of their fathers. A wild animal learns which path leads to safety or danger, and which strategy leads to success or distress. Many wild animals raised in captivity do not learn the skills required for survival and cannot be released into the wild without special effort. *Born Free*, the bestselling book by Joy Adamson and the Academy Award–winning film of the same name, describes the true story of a lion cub named Elsa, raised in captivity and the harrowing difficulties of releasing this lion into the wild. Elsa was severely injured by wild animals and nearly starved before learning how to live in the wild without human aid.[11]

...Salmon from Behavioral to Molecular Biological Approaches," *General and Comparative Endocrinology* 170, no. 2 (January 2011): 222–32, https://doi.org10.1016/j.ygcen.2010.02.003.

10 Navigation skills of migratory species rely on a combination of environmental cues, such as the Earth's magnetic field, the position of the sun and stars, and chemical signals. Memory also plays a significant role, especially in experienced migratory birds.

11 Joy Adamson, *Born Free: A Lioness of Two Worlds* (Pantheon Books, 1960).

The fox has entered our language as the epitome of guile and cleverness. Though the fox operates on the perceptual level of intelligence, it has "outfoxed" many a farmer over the centuries, stealing chickens and other small animals from farmers' pens, evading capture, and frustrating the farmer's efforts to keep his animals safe.

More recently, an "arms race" has developed between bears and the US National Park Service, which operates wilderness campgrounds with bears all around. The bears smell food the campers bring and will eat it if they can. To prevent bears from stealing the campers' food, the park service has installed "bear-proof" metal boxes at campsites. The idea is to have a latch on the box that can be opened by a camper using written instructions, but is too complicated for the bears to figure out. Yet each time the park service installs a new model, the bears use their problem-solving skills to open the bear boxes.[12] If you can design a bear box that can be opened by a camper arriving at a wilderness campsite and yet outsmart the bears' ability to open it, I'm sure the US National Park Service would like to talk to you!

The evolutionary benefit of consciousness is volitional awareness—the ability of an animal to focus attention on specific aspects of its environment to further its survival. Volitional awareness enabled perceptual intelligence to develop in advanced animals, where the animal chooses the best path in search of food or to flee from predators. A component of this intelligence is knowledge, both innate and learned. An animal uses its knowledge to choose the path that supports its life rather than one that destroys it. Otherwise, paraphrasing Pierson and Trout, what is knowledge for?

---

12   Lisa W. Foderaro, "Bears in the Adirondacks Defeat BearVault Food-Protection Container," *New York Times*, July 25, 2009, https://www.nytimes.com/2009/07/25/nyregion/25bear.html?hp.

Volitional awareness in man led to our ability to focus on thoughts. Concepts are the result, but concepts, unlike percepts, are not perceived directly. Every concept must be formed separately in each individual's human mind. First-level concepts are formed from direct experience, while high-level concepts are formed by combining other concepts. Let's take a deeper look at how concepts and conceptual intelligence is formed.

TABULA RASA

Every parent knows not to place a baby on a couch and walk away. Yet behavioral psychologists have attempted to show that babies are born with an innate fear of heights. They did this using a poorly constructed experiment in which crawling babies were placed on a table in the middle of which a glass surface was inserted to provide a perceptual cliff. The experimenter observed whether a baby would crawl across the transparent section to reach his beckoning parent on the other side of the table. Since babies do a lot of random crawling, it is impossible from this data to know whether the baby avoids the perceptual cliff due to an innate fear of falling or just due to chance. The behavioral psychologists claimed it was the former.

A better approach to testing a baby's fear of cliffs was developed by Karen Adolph.[13] Adolph built a variable-slope ramp to test whether babies differentiate between a ramp with a gradual slope or one with a steep slope. It turns out babies make no distinction. Babies attempted to go down steep ramps just as readily as they attempted to go down gradual ramps. (Parents were there to catch the tumbling babies.) Infants are wonderful experimenters; they

13   Karen E. Adolph, "Specificity of Learning: Why Infants Fall Over a Veritable Cliff," *Psychological Science* 11, no. 4 (July 2000): 290–95, https://doi.org/10.1111/1467-9280.00258.

quickly learn which slopes are possible and which are impossible to navigate. They would hesitate if a ramp was in the gray zone between the possible- and impossible-to-navigate slopes.

Interestingly, the "too steep to crawl down" lesson had to be relearned when the babies became toddlers. Walking is a skill with new limits that needs to be learned through trial and error by every human being.

Humans are born without innate ideas, not even the concept that you will fall if you step off a cliff.[14] All concepts must be formed separately through human effort in each person's mind. Aristotle expressed this well: We are born *tabula rasa*.[15]

Ayn Rand considered volitional consciousness to be man's indispensable characteristic:

Man is a being of volitional consciousness.[16]

This is true but must be reframed in terms of the Pierson-Trout hypothesis. All conscious beings have some degree of volition. However, an animal's volitional control is limited to where it directs its attention; it has no control over *what* it perceives, only where it looks. The same is true for man on the perceptual level—we have the ability to direct our gaze to whatever interests us, but we have no choice about what we see. How we interpret what we see, that's another matter entirely. Concept formation is entirely volitional. When Ayn Rand says, "man is a being of volitional consciousness," she means conceptual knowledge is entirely volitional. Each person must focus his or her mind to form every single concept in it. No

---

14 Babies are born with instincts to suckle, to cry, to grasp, and others, but none of this requires concept formation.

15 Aristotle, *De Anima*, 429.

16 FNI, 120.

automatic route to conceptual knowledge exists. Every idea we form is an act of will:

> Man's consciousness shares with animals the first two stages of its development: sensations and perceptions; but it is the third state, *conceptions*, that makes him man. Sensations are integrated into perceptions automatically, by the brain of a man or of an animal. But to integrate perceptions into conceptions by a process of abstraction, is a feat that man alone has the power to perform— and he has to perform it *by choice*. The process of abstraction, and of concept-formation is a process of reason, of *thought*; it is not automatic nor instinctive nor involuntary nor infallible.[17]

Percepts are conscious representations of particular things. Concepts are generalizations of all existents of a certain type.

> There is nothing automatic about concept formation. A man must focus his mind on the subject at hand to form a concept.

THE ANSWER

What is the relationship between my mind and the world? Man's mind has three different relationships with the world. The first relationship is subconscious and controls physiological functions, such as your heart rate, and automated actions, such as breathing and walking. You do not have direct control over your subconscious, although breathing and walking can be controlled at will by your conscious mind. The second relationship is perceptual. For example, when you are driving a car, you control where you focus your eyes, but you

17   FNI, 14.

cannot control what you see. The third relationship is conceptual. You must focus your mind on issues and entities in existence and form abstractions from what you perceive. The conceptual process is entirely voluntary. A concept cannot be formed in your mind unless you think about it.

IS SCIENCE OBJECTIVE?

SCIENCE IS A MODERN INVENTION

In medieval times, an intolerant philosophy claimed that truth emanated from God and His imagined thoughts. For over a thousand years, blasphemy, an utterance that questioned the role and power of God, was punished by torture and death. Yet, starting around 1400 CE, the modern world emerged. Truth in the modern world was radically different from God's truth derived from mystical revelations. In modern philosophy, truth was established by correspondence with reality. How did the world transition from faith to empiricism?

Religion's fall happened through stealth. Francis Bacon argued that examining and understanding God's creation would bring men closer to God. Bacon claimed God's truth could be found through inductive logic. The Baconian method, a precursor to the scientific method we know today, isolated the cause of a natural phenomenon. Bacon argued that careful, systematic observations of reality are necessary to produce quality facts, that it is a mistake to generalize beyond what the facts demonstrate, and one should gather additional data, perform experiments, and build up knowledge in a stepwise fashion.[1] While appearing to respect God, Bacon's empirical approach to "natural philosophy" made God irrelevant.

Bacon's *Novum Organum*, or "new logic," published in 1620, dealt medievalism a heavy blow. [2] The "old logic" was deductive, based on Aristotle's organum: the *Categories*, *On Interpretation*, the *Prior Analytics*, the *Posterior Analytics*, the *Topics*, and *On Sophistical Refutations*.[3] Deduction assumes a general truth and derives specific facts from this assumption. Bacon reversed this process. He argued that one should examine many examples of specific events and derive general truths from these examples. The process of deriving a general truth from specific cases is called *induction*.[4] By focusing on experiments in this world, Bacon established reality as the arbiter of truth.

In Galileo Galilei and Isaac Newton's capable hands, Bacon's approach blossomed into the scientific method. Both Galileo and Newton began with a particular effect observed in nature, formed a hypothesis of the cause of this effect, performed experiments to prove or disprove the hypothesis, and refined and repeated the experiments and the observations until a general truth was formed.

---

1   Mary B. Hesse, "Francis Bacon's Philosophy of Science," in *A Critical History of Western Philosophy*, ed. D. J. O'Connor (Free Press, 1964), 141–52.

2   Francis Bacon, *The* Instauratio Magna *Part II:* Novum Organum *and Associated Texts*, ed. Graham Rees (Clarenden Press, 2004).

3   Robin Smith, "Aristotle's Logic," *The Stanford Encyclopedia of Philosophy*, winter 2022 ed., ed. Edward N. Zalta and Uri Nodelman, https://plato.stanford.edu/ENTRIES/aristotle-logic/.

4   Bacon did not entirely understand the mechanism by which truth is obtained: Truth is established through identification, not induction, as explained in *Question 8—Does induction prove anything?*

In Galileo's case, he showed the period of a pendulum depended on its length but not length of the arc it swings,[5] that in the absence of wind resistance, all objects fall at the same accelerating rate, and, using a telescope, that Jupiter has moons that orbit the planet. Newton's first scientific endeavor was the study of light, performing experiment after experiment to unlock light's secrets; then Newton moved on to mechanics, beginning with Galileo's experiments and observations but adding many of his own; and, finally, Newton hypothesized that a gravitational field emanates from all bodies containing mass and showed that the orbits thus derived agreed with Tycho Brahe's observations of planetary motion and Kepler's laws. The young Newton also invented calculus but did not publish this work until much later.

Newton's pioneering book, *Philosophiæ Naturalis Principia Mathematica (Mathematical Principles of Natural Philosophy)*, published in 1687, turned men from looking for revelations from God to seeking truth by observing and examining reality.[6]

Alexander Pope captured the new spirit with his couplet:

Nature and Nature's laws lay hid in night:
God said, Let Newton be! and all was light.[7]

God became remote. According to prevailing philosophies in the Age of Enlightenment, God set the world in motion but did not interfere with its day-to-day operation. Men were allowed to examine

---

5   This is true only with small-angle arcs.
6   Isaac Newton, *Philosophiæ Naturalis Principia Mathematica* (S. Pepys, 1687).
7   Alexander Pope, "Epigram on Sir Isaac Newton," in *A Random Walk in Science*, ed. Robert L. Weber and Eric Mendoza (The Institute of Physics, 1973), 59.

reality, to understand it, and to live better for it. John Locke invented the concept of individual rights—that each man has a right to life, liberty, and property—and man was set free.[8]

These three—Bacon, Locke, and Newton—were the heroes of the Enlightenment. Thomas Jefferson wrote:

> Bacon, Locke and Newton,...I consider them as the three greatest men that have ever lived, without any exception, and as having laid the foundation of those superstructures which have been raised in the Physical and Moral sciences.[9]

Jefferson borrowed from John Locke's thoughts to write in the American Declaration of Independence:

> We hold these truths to be self-evident, that all men are created equal, that they are endowed by their Creator with certain unalienable rights, that among these are Life, Liberty and the pursuit of Happiness.

The medieval world became modern. Freed from the overwhelming oppression of state and religion, people became richer, science flourished, and life bloomed. The average hourly earnings of Americans increased from $0.10 in 1860[10] to $34.75 in 2024;[11] life

8   John Locke, *Two Treatises on Government*, ed. Peter Laslett (Cambridge University Press, 1988), sec. 87, 123, 209, 222.
9   Thomas Jefferson to John Trumbull, February 15, 1789, https://founders.archives.gov/documents/Jefferson/01-14-02-0321.
10 National Bureau of Economic Research, "Laborers' Average Hourly Rate of Wages, Unweighted for United States," retrieved from FRED, Federal Reserve Bank of St. Louis, accessed December 5, 2024, https://fred.stlouisfed.org/series/A08138USA052NNBR.
11 US Bureau of Labor Statistics, "Average Hourly Earnings of All Employees,

expectancy increased from forty years in 1860 to almost eighty years in 2020;[12] and people went from heating their homes using fireplaces to comfortable central heating systems, and from driving around in horse-pulled buggies to luxurious automobiles. The drudgery of medieval times was replaced by the American dream—a life of comfort and plenty to those willing and able to work for it.

THE POSTMODERN ATTACK ON SCIENCE

Yet the transformation from medieval to modern was not as strong as it seemed: The Enlightenment was undercut by remnants of medieval philosophy. Francis Bacon, the inventor of the scientific method, proposed observation as an aid to understanding God; Isaac Newton, the hero who gave us "natural law," dabbled for decades in alchemy and religious mysticism;[13] and John Locke, that champion of individual rights, urged the authorities to persecute atheists, because he argued that the Bible was consistent with reason.[14] No "existence first/consciousness second" philosopher appeared until Ayn Rand formulated objectivism in the mid-twentieth century. Absent a foundation in existence, science was vulnerable to attack.

Jean-François Lyotard, the French philosopher who coined the term *postmodernism*, aimed to topple science from its lofty perch:

I define postmodern as incredulity toward metanarratives. This incredulity is undoubtedly a product of progress in the sciences:

Total Private," retrieved from FRED, Federal Reserve Bank of St. Louis, accessed December 5, 2024, https://fred.stlouisfed.org/series/CES0500000003.

12   Aaron O'Neill, "Life Expectancy in the United States, 1860–2020," Statista, August 9, 2024, https://www.statista.com/statistics/1040079/life-expectancy-united-states-all-time/.

13   Gale E. Christianson, *Isaac Newton* (Oxford University Press, 2005).

14   John Locke, *The Reasonableness of Christianity* (1695).

but that progress in turn presupposes it...if a metanarrative implying a philosophy of history is used to legitimate knowledge, questions are raised concerning the validity of the institutions governing the social bond...[15]

Postmodernism attacks science at its root. Lyotard claims that science does not examine reality, that it is merely a "metanarrative," a story people tell themselves. Science, in this view, is no more real than other possible "metanarratives." Further, using science to "legitimate knowledge" is harmful, according to Lyotard, because it delegitimizes "the institutions governing the social bond." Science frees each person to think for himself or herself, while postmodern philosophers prefer a society where social institutions (i.e., the state) dictate the rules.

Metaphysics or metanarrative? An odd Ultimate Question. Does science reveal the truth about the nature of the universe, or is science just a story we tell ourselves? To find out, let's look at the roots of postmodern philosophy.

THE POSTMODERN ATTACK ON OBJECTIVITY

Philosophy is divided into four movements: the Greco-Roman period from 600 BCE to 400 CE, the medieval era between 400 CE to 1400 CE, the modern period from 1400 CE to the present, and the postmodern advance from the late 1700s to the present. The transitions from Greco-Roman to medieval, and from medieval to modern took centuries. The battle between modernism and postmodernism has been ongoing for roughly 250 years. Postmodernism has not yet

---

15   Jean-François Lyotard, *The Postmodern Condition: A Report on Knowledge*, trans. Geoff Bennington and Brian Massumi (University of Minnesota Press, 1984).

won, but postmodern philosophy dominates academia around the world today.

Postmodernism's rise is due to a quirk in history. Modern philosophy is a mirage—a philosophy of scattered ideas without a unifying tenet, thought, or theme. When one queries "modern philosophy" in the Stanford Encyclopedia of Philosophy, not a single entry appears. In contrast, entering "postmodern philosophy" generates dozens of articles. Postmodern philosophers can safely ignore modern philosophy because its precepts are undefined.

Yet there is a philosophy that strengthens modernism to make it impenetrable. In 1957, Ayn Rand published *Atlas Shrugged* and introduced objectivism, which transforms modern philosophy into a unified whole. Rand took the elements of modern thought that made sense and built a structure to support them.

Objectivism is modernism made bulletproof.

A good way to distinguish between the different philosophies is *The Philosophy Chart* below, which provides an essential guide. Originally created by Stephen Hicks and slightly modified here, the chart presents the principal characteristics of the medieval, modern/ objective, and postmodern periods. The Ultimate Answers for metaphysics, epistemology, ethics, human nature, and politics during each period is identified in The Philosophy Chart.

If knowledge is objective, then truth is derived by forming concepts that correspond to things that exist.

In this case, each individual mind is the ultimate arbiter of truth— each mind can examine the concept and the data from which it is derived and determine whether the concept is true or false. If knowledge is a metanarrative formed by society, then determining the truth of a statement is a group activity. In a metanarrative, the group decides what is true and what is false. In the first case, the individual is supreme and individual rights and capitalism follow. In the second case, the group determines your fate—you have no right to defy society's whim. In the first case, you are free to do as you please as long as you do not initiate the use of force against others. In the second case, society is your master, socialism your fate.

The Philosophy Chart[16]			
	Medievalism	**Modernism/ Objectivism**	**Postmodernism**
Metaphysics	Realism: Supernaturalism	Realism: Naturalism	Anti-realism
Epistemology	Mysticism and Faith	Objectivism: Experience and reason	Social subjectivism
Ethics	Collectivism: Altruism	Individualism: Self-interest	Collectivism: Egalitarianism
Human Nature	Original sin: Subject to God's will	Tabula Rasa: Autonomy	Social construction and conflict
Politics	Feudalism	Capitalism	Socialism
When and Where	Medieval Europe, Middle East, and Asia	The Enlightenment: Twentieth and twenty-first century sciences, engineering, and business	Twentieth and twenty-first century humanities and related professions

16   Stephen R. C. Hicks, *Explaining Postmodernism: Skepticism and Socialism from Rousseau to Foucault* (Expanded Edition) (Ockham's Razor Publishing, 2011), 15. Slightly modified.

In objectivism, science is objective. Science is a systematic endeavor that builds and organizes knowledge in the form of testable explanations and predictions about the universe.[17]

To destroy objective thinking, Jean-François Lyotard, the person who originated the term *postmodernism*, found that he had to destroy science. He did this by calling science a "game" with rules of its own making:

> Science has always been in conflict with narratives. Judged by the yardstick of science, the majority of them prove to be fables. But to the extent that science does not restrict itself to stating useful regularities and seeks the truth, it is obliged to legitimate the rules of its own game. It then produces a discourse of legitimation with respect to its own status, a discourse called philosophy. I will use the term modern to designate any science that legitimates itself with reference to a metadiscourse of this kind making an explicit appeal to some grand narrative...[18]

The battle between objectivism and postmodernism is a battle over metaphysics and epistemology. Does nature exist independently of man's thoughts, or do man's thoughts define nature? What steps must be used to ensure a concept correctly describes that which exists? Conflicting ideas about the nature of man, of morality, of politics, and of economics are impossible to resolve without agreement about the method by which we can determine truth from falsehood.

Objectivism's metaphysics is simple: Nature exists independently of man's thoughts and wishes. Nature is what it is.

---

17   Edward O. Wilson, "The Natural Sciences," in *Consilience: The Unity of Knowledge* (Vintage, 1999), 49–71.
18   Lyotard, *The Postmodern Condition*, 1.

Objectivism's epistemology is based on reason. Man will never know everything, but he can validate ideas about things that exist by studying nature and applying the scientific method. Our knowledge of existents is vastly more advanced today than it was in the medieval era.

Postmodernism's metaphysics—that true reality is unknowable—dismisses metaphysics in favor of epistemology. Reality may or may not exist but who cares, say the postmodernists, it is unknowable anyway. In the postmodern narrative, what counts is consensus, not correspondence with reality.

THE ANSWER

Is science objective? Science is objective in a galaxy with objective rules, a galaxy that exists independently of thoughts, where consciousness is awareness of existents, not their creation. This is the objectivist's view. In the postmodern view, science is just a story we tell ourselves. In this case, scientists are no better than witch doctors, making incantations they do not understand, with truth determined by the group's thoughts rather than existence. The choice is yours, but it is hard to imagine the progress we have seen in medicine, biology, chemistry, physics, and engineering over the past four centuries if science was just a story we tell ourselves rather than an explanation of how reality actually works.

ARE SCIENTIFIC LAWS CERTAIN?

THE ABSOLUTE TRUTH

Sometimes a contradiction is hard to spot.[1] But the contradiction in this statement, "There are no absolutes!," is simple. The speaker is declaring with absolute certainty that absolute certainty does not exist. The contradiction is self-evident: We can be absolutely certain the statement, "There are no absolutes!," is false.

Contradictions do not exist because there is one universe. Everything that exists has a nature whether you identify that nature

---

[1] The barber paradox is one such case: A barber, who lives on an island, shaves all those men who live on the island who do not shave themselves, and only those men. It is impossible for the barber to shave himself, or not to shave himself. If the barber shaves himself then he is a man on the island who shaves himself hence he, the barber, does not shave himself. If the barber does not shave himself then he is a man on the island who does not shave himself hence he, the barber, shaves him(self).

correctly or not. The issue is the correctness of your identifications, not whether things exist.

> Provided you identify things correctly, your
> ideas will correspond to existents.

Existents do not contradict themselves; they simply are.

> Reason is the faculty that identifies and integrates the material provided by man's senses.[2]

Sense data comes in, and the mind forms concepts from this data and integrates these concepts into a unified whole. If you discover a contradiction in your thinking, the error is in your thinking, not in existence. Knowledge is validated by using the principles of logic, "the art of non-contradictory identification."[3] Knowledge requires a strict adherence to the facts of experience and an understanding of the idea: Contradictions do not exist.

It is easy to string words together so they do not make sense, such as: "This sentence is false." After you eliminate semantic errors, the issue remains: How do you know your identifications are correct? Can you be certain your thinking is true?

Each individual's knowledge is a growing sum, from infancy, to youth, to adulthood. As a person's knowledge grows, he or she must integrate new ideas with preexisting ideas to form a unified whole. At no stage will an individual, or even all the individuals on earth put together, know everything that exists. Learning is never finished because existence is limitless. In Leonard Peikoff's words:

---

2   VOS, 10.
3   FNI, 125.

Metaphysically, there is only one universe. This means that everything in reality is interconnected. Every entity is related in some way to the others; each somehow affects and is affected by the others. Nothing is a completely isolated fact, without causes or effects; no aspect of the total can exist ultimately apart from the total. Knowledge, therefore, which seeks to grasp reality, must also be a total; its elements must be interconnected to form a unified whole reflecting the whole which is the universe.[4]

Physics is the study of this one universe. But this one universe has many layers. This layering of information is a central aspect of knowledge. Concept formation is like a set of Russian Matryoshka dolls, with one concept imbedded in another. Each doll must fit, or else the whole system bursts apart.

All knowledge is contextual in an interconnected world.

A person can know some relationships but not all relationships. A child begins by knowing that things exist: "I see Mommy. She is speaking into a box she is holding in her hand. The box is speaking back to her." Later, as the child grows, he or she learns that the box is called a phone, that it is Daddy's voice on the phone, but that Daddy is far away. Still later, he or she learns the box is called a smartphone, it has a screen, and the child learns he or she can play games on it. Growing into adulthood, the child learns the smartphone contains electronic circuitry, that this circuitry converts the sound waves produced by the speaker's voice into electronic signals, that these signals are transmitted using electromagnetic waves and, after several steps, arrive at the receiver where they are converted back into sound waves.

---

4   OPAR, 123.

At no stage in this process is the child wrong. His or her mommy exists, the box exists, and the box produces audible sound waves. Later, when the child learns the box is a game station as well as a phone, it adds to his or her knowledge but does not contradict it. Still later, when the child learns that, in a phone conversation, the words coming out of the phone originate from Daddy who is far away, it does not contradict the toddler's thought, *The box is speaking back to her*. A human voice was coming out of the box; within the context of a toddler's level of knowledge, this is a correct statement. Learning that Daddy is far away does not change this fact.

Leonard Peikoff writes:

> All thought, argument, proof, refutation
> must start with that which exists.[5]

Thought is hierarchical, beginning with the absolutism of existence, followed by primary identifications of existents, and proceeding to higher-level concepts. Certainty in thought is possible only if you specify its location in this hierarchy. All knowledge is contextual:

> *Certainty is a contextual assessment*, and in countless situations the context permits no other. Despite the claims of skeptics, doubt is not the human fate, with cognition being an unattainable ideal. Doubt, rationally exercised, is a temporary, transitional state, which is applicable only to (some) higher-level questions—and which itself expresses a cognitive judgment: that the evidence one has is still inconclusive. As such, doubt is made possible only by a vast context of knowledge in the doubter's mind. The doubter must know both facts and logic; he must know the facts known so far—and also the means by which in principle his doubt

5   OPAR, 168.

is eventually to be removed, i.e., what else is required to reach full proof?[6] (italics added)

THE ANSWER

Are scientific laws certain? Scientific laws are based on the proper identification of existents. Once identified, you can be certain that an existent will behave according to its nature. You do not have to observe this nature over and over again to know that it is true. Keep in mind, however, you will never know *everything* about anything. Certainty is contextual. Knowledge is limited by the context of how the existent is identified.

6   OPAR, 181.

DOES INDUCTION PROVE ANYTHING?

THE PROBLEM OF INDUCTION

With respect to certainty, Karl Popper was the ultimate skeptic. He argued that since no number of experiments could ever *prove* a scientific theory but a single experiment could *disprove* one, science should be based on falsification.[1] You can never be certain, in Popper's view, because someone may falsify your theory at any time.

Popper's error, and the error made by philosophers through the centuries, is to assume that scientific theories are proved through induction. This is not the case.

---

1   Karl Raimund Popper, *The Logic of Scientific Discovery* (Springer, 1935; repr., Routledge Classics, 2002).

Certainty in science is established through the Identity Axiom. Induction has nothing to do with it.

The "problem of induction" was first formulated by David Hume in 1739.[2] Hume argued that seeing past instances of cause and effect did not necessarily imply that the same cause would have the same effect in the future. For example, men have observed gunpowder touched by a flame explode thousands of times in the past, but how do you know that it will explode the next time you see a flame strike gunpowder? More generally, how do you infer a general truth from any number of specific instances of an event that happened in the past?

The answer, of course, is in properly identifying the existents involved. In the case of gunpowder, assuming the chemical elements are identified properly, the ignition temperature is 495 K.[3] The explosion is in the nature of the chemical elements, not in how many times you observe it.

Hume's "problem of induction" is caused by a primacy of consciousness worldview. This error occurs when men place their thoughts above existence.[4] Everything that exists has a nature. Induction supposes you must observe this nature repeatedly to make sure it repeats. And, even when it repeats, you cannot be sure that it will repeat once more. In this process, your observation is in the driver's seat, not existence.

A primacy of existence worldview has existence in the lead role. In Ayn Rand's words:

---

2   David Hume, *A Treatise of Human Nature* (1739).

3   Michael L. Hobbs and Michael J. Kaneshige, "Cookoff of Black Powder and Smokeless Powder," *Propellants, Explosives, Pyrotechnics* 46, no. 3 (January 2021): 484–93, https://doi.org/10.1002/prep.202000214.

4   ITOE, 53.

Existence is Identity, Consciousness is Identification.[5]

Man's role is to identify that which exists. The scientific method is a procedure for identifying existents, not a cog in a never-ending conceptual chain. Once an existent has been identified, it behaves according to its nature. The number of observations of this nature is irrelevant. The important point is to identify the existent—how it behaves is an attribute of the existent.

Consider the law of conservation of energy. This law states that, in a nonrelativistic system, the total energy of a system before an event is identical to the total energy of the system after the event: *Total energy before = Total energy after.*

As an example, suppose I hit a cue ball with 1 joule of energy. The cue ball rolls along the pool table and strikes the eight ball at a 45° angle. The cue ball and the eight ball fly apart at 90°, each one carrying ½ joule of energy. Since kinetic energy is ½ mass times velocity squared, the velocity of the cue ball and of the 8-ball is $1/\sqrt{2}$ times the velocity of the cue ball prior to impact. Energy is conserved through the impact.

A skeptic would argue my "conservation of energy" calculation ignored several effects. It did not take into account the friction of the ball rolling along the felt surface of the pool table, the air resistance of the moving ball, the rotational energy of each ball, the heat generated when the cue ball struck the eight ball, the deviation from 45° in hitting the eight ball, the slight difference in mass of the cue and eight balls, and others. Yet ignoring these minor effects does not invalidate the law of conservation of energy. It simply means nature has more layers to it than my calculation employed.

Conservation of energy is like accounting: You must keep track of each dollar in a profit and loss statement, but it is not an existential

---

5   FNI, 125.

crisis if you are off by a penny or two. At least it is not an existential crisis in engineering. Physicists will hunt down every last penny!

The law of conservation of energy is *testable* but not *falsifiable*. Provided all physical effects are included in the analysis, the only way energy is not conserved is with a miracle. Only God can accelerate a billiard ball *without any physical basis*. Popper's falsification argument is at odds with reality.

It is not possible to falsify a true statement.

"Proof" is not a numbers game. It matters not whether you have verified the law of conservation of energy one hundred times or a million times. What counts is you identify an existent properly and apply the Identity Axiom to this existent properly. The equal sign in the conservation of energy says the two sides of this equation are identical. The law of conservation of energy states the total energy of an isolated system remains constant. If the energy after an event does not equal the energy before the event, you need to find the discrepancy in your calculations, not argue that "a miracle happened."

The key phrase in the previous discussion is "identify an existent properly." A proper identification must satisfy three criteria:

1. The hypothesis of a new existent must not contradict the corpus of existing knowledge.
2. The hypothesis must explain observed phenomena within the limits of the observation.
3. The hypothesis is valid only within the context of the idea being proposed.

The first of these criteria derives from the nature of existence. As Peikoff said, there is only one universe, a universe in which everything is interconnected.[6] Everything interacts with other things in some way; nothing is isolated from all else in all ways and in all respects. The universe is the totality of all that exists. No concept can be isolated from all other concepts in all ways and in all respects. For this reason, all knowledge is interconnected. Knowledge is the totality of all correct thoughts. New knowledge can add to the corpus of existing knowledge but never contradict it. In Ayn Rand's words:

> No concept man forms is valid unless he integrates it without contradiction into the total sum of his knowledge.[7]

Knowledge here means thoughts that correspond to reality. Flights of fancy, such as "the Earth is the center of the universe," do not qualify as knowledge.

The second criteria is fundamental to the scientific method. The results derived from a hypothesis must agree with observations for the hypothesis to be valid.

The third criterion is best illustrated by the skeptic who says about the energy conservation law: "You're wrong. Einstein falsified the law of conservation of energy through his famous equation, $E = mc^2$. This equation says that energy can be converted into mass, and vice versa, so that in general mass-energy is conserved, not energy alone."

The skeptic's error is dropping the *context* of the scientific statement. Ayn Rand observed:

> Concepts are not and cannot be formed in a vacuum; they are formed in a context; the process of conceptualization consists of observing the differences and similarities of the existents *within the*

---

6   ITOE, 39.
7   FNI, 126.

field of one's awareness (and organizing them into concepts accordingly). From a child's grasp of the simplest concept integrating a group of perceptually given concretes, to a scientist's grasp of the most complex abstractions integrating long conceptual chains—all conceptualization is a contextual process; the context is the entire field of a mind's awareness or knowledge at any level of its cognitive development...the content of [an individual's] concepts is determined and dictated by the cognitive content of his mind, i.e., by his grasp of the facts of reality. If his grasp is non-contradictory, then even if the scope of his knowledge is modest and the content of his concepts is primitive, *it will not contradict the content of the same concepts in the mind of the most advanced scientists.*[8]

Einstein's modification of the conservation of energy law expanded our understanding of nature; it did not contradict it. Mass is converted into energy in the transmutation of elements, not in the collisions of billiard balls. Firing a neutron at an atomic nucleus may convert some of the mass into energy, but this is a different process from one billiard ball hitting another. Low velocity interactions, such as billiard balls impacting one another, are governed by the law of conservation of energy; high velocity interactions, such as atomic particle interactions, are governed by the law of conservation of mass-energy. Both laws are valid provided the context of the law's applicability is properly understood.

CERTAINTY IS POSSIBLE THROUGH IDENTIFICATION, NOT INDUCTION

The cases for the existence of mass, atoms, and space-time appear certain, but how do we know this? The traditional argument for

---

8   ITOE, 42–43.

certainty is through induction, the process of deriving a general truth from specific instances of events. The case for induction is laid out in *The Logical Leap: Induction in Physics* by David Harriman:

> First, every concept and every generalization contained within the theory must be derived from observations by a valid method...[9]

> Second, a proven theory must form an integrated whole...[10]

> The third criterion pertains to the range of data integrated by the theory. The scope of a proven theory must be determined by the data from which it is induced; i.e., the theory must be no broader or narrower than required to integrate the data...[11]

These three criteria are necessary for a generalization to be true, but are they sufficient to prove the theory? Harriman says it takes a logical leap:

> The three criteria describe the relationship between a proven theory and the evidence supporting it. When every aspect of the theory is induced from the data (not invented from imagination), and the theory forms a cognitive whole (not an independent collection of laws), and the scope of the theory is objectively derived from the range of data—then the theory truly is an integration (criterion 2) of the data (criterion 1), no more and no less (criterion 3).[12]

---

9 David Harriman, *The Logical Leap: Introduction to Physics* (New American Library, 2010), 184.
10 Harriman, *The Logical Leap*, 184.
11 Harriman, *The Logical Leap*, 185.
12 Harriman, *The Logical Leap*, 186.

Except that this argument for "proof" relies on mental processes: (1) generalizations derived from observations; (2) mental integration of the theory into a whole; and (3) the range of data used in the mental integration. Since mental processes are volitional, they are subject to error. No argument that relies purely on mental processes can be certain.

Existence exists, and only existence exists. Consciousness is the process of identifying that which exists. Hence, certainty is possible only through identification. Just as a young child knows a cell phone exists and learns more and more about its operation as he or she grows, a scientist must begin with a basic definition of something that exists and study it. Certainty requires we focus on existents and the process by which existents are identified.

Mass was the first indirectly observed existent to be identified. People had experienced force and observed acceleration prior to Isaac Newton, but Newton was the first to explain that a quantity— mass—connects the two. Mass exists and is just as real as force and acceleration, but it lay hidden from the thoughts of ordinary men. Newton's genius opened the world to a new worldview in which mass plays a central role.

A universe without mass is impossible to imagine. Without mass, objects would accelerate without cause, raindrops would not fall from the sky, and the universe would dissolve into nonscientific chaos. Life and consciousness would not be possible. That much is certain.

Now imagine a universe without atoms. Without atoms, chemical reactions could not take place, fire could not happen, air would not be composed of bouncing molecules, and DNA would not exist. Life and consciousness could not exist without atoms.

Similar arguments can be made regarding elementary particles and space-time. Without elementary particles and space-time, the sun would not shine, light could not propagate, and atoms could not be formed. Life as we know it would cease to exist.

Certainty is contextual but existence is not. Existence exists, or else there is nothing to talk about. You exist, or else there is no point in talking. And mass, atoms, elementary particles, and space-time exist—otherwise life, consciousness, and the universe could not exist. Of that, we can be certain.

THE ANSWER

Does induction prove anything? No, induction requires a chain of never-ending observations to prove anything, an existential impossibility.[13] Proof is achieved through identification. Observations are required to identify things, but once an existent has been identified correctly, it will behave according to its nature. Identification must be placed in context, however. Man can know some aspects of an existent but not all aspects of it. Certainty is possible only within the context of the identifications made.

---

13   Induction in mathematics is different from induction in philosophy. To prove a formula for the sum of a mathematical series, one shows that the formula is true at the (n + 1)'st step provided it is true at the n'th step. In this case, examining two instances of the formula is equivalent to examining an infinite number of instances.

WHERE DOES KNOWLEDGE START?

THE SCIENCE OF EXISTENCE

Testability versus falsification highlights the difference between objectivism and postmodernism. Testability implies that man can understand the nature of reality and have testable ideas about how nature behaves. Falsification implies our knowledge of reality is never certain, and our ideas are one experiment away from being a myth. In objectivism, our knowledge grows with each new discovery. In postmodernism, science is a game played with human-invented rules. Objectivism ties science to reality. Postmodernism sets science adrift.

Ayn Rand built objectivism on a set of axioms. She defined a philosophic axiom as:

...the identification of a primary fact of reality, which cannot be analyzed, i.e., reduced to other facts or broken into component parts. It is implicit in all facts and in all knowledge. It is the fundamentally given and directly perceived or experienced, which requires no proof or explanation, but on which all proofs and explanations rest.[1]

In other words, a philosophic axiom is given ostensively.

The primary axiomatic concept is *existence exists*.

You experience existence directly. You open your eyes and there it is. The proof of existence is all around. This proof is not deductive. Deductive proof requires logic, and logic is derived from existence: If nothing exists, there is no logic; if logic exists, there must be something to be logical about.

The second philosophic axiom is *consciousness exists*.

Consciousness, like existence, is self-evident. Any attempt to deny consciousness by saying "consciousness does not exist" must use consciousness in its denial. Logic and explanation presuppose consciousness to be meaningful.

The axioms, *existence exists* and *consciousness exists*, are not only compatible with science, they are the bedrock on which science rests. *Existence exists* and *consciousness exists* are testable phenomena, science's distinguishing characteristic compared to unscientific speculation. We test the validity of these two axioms every single second of every waking day. Without existence, there is nothing to study; without consciousness, there is no possibility of study. Science begins with the two axioms: *existence exists*, and *consciousness exists*.

---

1   ITOE, 55.

The third philosophic axiom is *identity*.

This is a more subtle philosophic axiom than the other two. The Law of Identity is attributed to Aristotle: A thing is what it is; A is A. But why is this axiomatic? Why is identity implicit in all facts and all knowledge?

Ayn Rand's answer is that everything that exists, every existent, has a specific nature with specific characteristics. An existent apart from its characteristics is removed from its identity; an existent without an identity is nothing and nothing does not exist.

There is a lot to unpack here. The first point to note is the word "nothing" in the above statement is different from the "nothing" in common use. If you say, "There is nothing in the refrigerator," you mean, "There is no food in the refrigerator." Of course, there is space in the refrigerator, there are shelves, there is air, there is light when the door is open, but there is nothing to eat or drink. Space, shelves, air, and light are all assumed existents not relevant to the conversation of wanting to eat food. Similarly, if you say, "Nothing's the matter," you mean, "While there are issues, I can handle them." The "nothing" in the above sentence is an evasion, minimizing the problems and conflicts that prompted the comment.

The "nothing" outside of existence is not like an empty refrigerator or a person minimizing issues. Space, shelves, air, light, problems, and conflicts are all part of existence. The alternative to existence is not a void where something is missing in the universe. There is no universe, no space, time, issues, ideas, or thoughts without existence. Anything you know is ruled out if existence is ruled out. The alternative to existence is "nothing" and "nothing" has no attributes.

The statement, "an existent without an identity is nothing and nothing does not exist," now begins to make sense. If you have an entity in existence, it has attributes, and these attributes define its identity. Without attributes, an existent has no identity and is "nothing."

Because we live in existence,
the alternative, "nothing," does not exist.

Identity provides the link between existents and knowledge; knowledge is impossible if existents have no identity. Ayn Rand put it pithily: "Existence is Identity, Consciousness is Identification."[2]

Stones exist with or without the awareness of any conscious being. When you hold a stone in your hand, it is small, hard, and dense. These and other attributes must exist for you to identify it. If you remove all the stone's attributes, you have nothing left.

Identification occurs in the brains of conscious beings. A stone in nature simply is. The knowledge that it exists, its identification, is a result of conscious awareness.[3] Knowledge is the mapping of existents into thoughts that allow a conscious being to interact with existents purposefully. This knowledge is neither intrinsic (imbedded in the object itself) nor subjective (created in conscious brains entirely from whole cloth). Knowledge is objective: It is the mapping of the attributes of existents into conscious thoughts.

The relationship between thoughts and brains is analogous to the relationship between software and hardware in a computer system. Neither software nor thoughts can be touched, but the software in a robot and the thoughts in a man determine how the robot moves and how the man behaves. Software is nonmaterial but determines what the computer computes. Consciousness is nonmaterial but determines how an animal or a man acts.

---

2   FNI, 125.

3   Animals are aware of stones and other objects on a perceptual level. Men are aware of such things on the conceptual level.

THE PHILOSOPHIC COROLLARIES

Ayn Rand wrote that "identity" is a corollary of "existence."[4] According to Peikoff, "A corollary of an axiom is not itself an axiom; it is not self-evident apart from the principle(s) at its root (an axiom, by contrast, does not depend on an antecedent context). Nor is a corollary a theorem: It does not permit or require proof; like an axiom, it *is* self-evident (once its context has been grasped)."[5] The Law of Identity is not as readily apparent as the axiom of existence. Existence hits you in the face, but it takes thinking to realize that things are what they are and that you have no choice about it. This lack of choice *is* self-evident. Only your identification of the nature of the thing you examine is up in the air.

Existence includes everything that exists, and *consciousness* includes all thought processes; other axiomatic concepts are necessarily corollaries of the first two. We have already discussed how the third primary axiom, *identity*, is a corollary of existence. Let us now examine the other philosophic corollaries embedded in the first two.[6]

We begin with *space* and *time*.

Space and time are part of existence, are fundamentally given, are directly perceived, require no proof or explanation, but all proofs and explanations are impossible without them.

---

4 ITOE, 55.
5 OPAR, 15.
6 Leonard Peikoff identified three secondary axioms: "The concept of 'entity' is an axiomatic concept" (OPAR, 12); "The validity of the senses is an axiom" (OPAR, 39); and "The principle of volition is a philosophic axiom" (OPAR, 70). Harry Binswanger wrote: "...sensory awareness is axiomatic..." (HWK, 59); "...volition is an axiom of the conceptual level" (HWK, 355); "a few other axiomatic concepts [exist], the most noteworthy being 'entity' and 'action' or 'change'" (HWK, 167).

Space is not a relationship. It is the fabric of existence. Saying, "There's no such thing as grandfathers" provides an inconsistency in logic; saying, "There's no such thing as space" denies existence as such. We perceive space and time directly. It is impossible to conceive of any material object without it being embedded in space. It is impossible to think without experiencing the flow of time.

Space and time include *entities* and *actions*. Again, these existents are fundamentally given, are directly perceived, and require no proof or explanation. Entities and action cannot exist without space and time but do exist independently of consciousness.

Consciousness includes the axiomatic concept *self*.

> Self is axiomatic since it is self-evident, and any attempt
> to deny it, "I have no self," is self-contradictory.

Concepts are formed only in the individual human brain. No one can be conscious or have knowledge apart from himself or herself.

Self depends on *sense perception* and the *validity of the senses*. Awareness is impossible without sense perception; knowledge is impossible without the validity of the senses.

Identity depends on *volition* since a consciousness must focus to identify facts of reality. Truth is determined through choice, right or wrong: Are the facts of reality identified correctly?

> Thinking is an act of choice.

Leonard Peikoff identifies "...causality...as a corollary of identity."[7] However, Ayn Rand writes:

---

7   OPAR, 15.

The law of causality is the law of identity applied to action.[8]

Thus, causality is a logical consequence of the axioms "identity" and "action." The consequence of an event is determined by the entity's identity and the action it undergoes. For this reason, causality is not axiomatic.

HISTORICAL CONTEXT

Evolution is existence based and is compatible with a metaphysics where existence exists, and consciousness evolved to perceive existence. However, Ayn Rand first presented her metaphysics in her novel *Atlas Shrugged* in 1957 and further developed her ideas in nonfiction works written in the 1960s and 1970s.

When one enters "philosophic axioms" into the Stanford Encyclopedia of Philosophy search engine, random entries come up, primarily axioms concerning mathematics and logic. Euclid presented an axiomatic structure for geometry in his *Elements* written circa 300 BCE, a structure that remains with a few changes to the present day. Axioms have been the stated basis of mathematics and logic ever since. Given this, it is strange that Ayn Rand's axiomatic approach to metaphysics, an approach that is correct on the merits, is rarely cited today.

Perhaps the issue is a confusion between logic and revelation. Ancient philosophers such as Plotinus employed axioms derived through revelation to establish the supremacy of "The One" (i.e., God). Medieval scholars treated God as a self-evident truth not to be questioned without a charge of heresy. Modern philosophers rightly condemn such fact-free approaches to knowledge; axioms defined by revelation are arbitrary and capricious.

---

8   AS, 1037.

That said, equating the axiom "existence exists" with religious revelation is a mistake of major proportions. The two approaches to truth are opposites: One accepts the world as it is and views consciousness as a means to understanding existence; the other places consciousness in the driver's seat, says arbitrary commandments cannot be questioned and argues that "real" reality is beyond man's reach.

Consciousness is a product of evolution, just like everything else concerning life on earth. Pierson and Trout argue that the evolutionary benefit of consciousness is volitional awareness, an awareness that enables volitional movement. Concept formation evolved later, as recently as 30,000 years ago. Ayn Rand explained that concept formation requires focus that is only possible through volitional control. All this is fact based and logical but runs into the buzz saw of determinism, the doctrine that everything is determined by the laws of physics. In answering *Question 36—What's the physics underlying free will?*, we show that determinists misinterpret physics. Nothing in physics implies the impossibility of volitional control. Thinking and understanding is what our species evolved to do.

THE ANSWER

Where does knowledge start? Knowledge starts with the axiomatic concepts of *existence, consciousness*, and *identity*. These axioms cannot be proved but all thought depends on them. Since logic exists, is available through consciousness, and has a defined nature, logic without existence, consciousness, and the Law of Identity is a contradiction in terms. Accepting existence, consciousness, and identity as the three primary concepts from which all else is derived is man's first step to understanding the galaxy.

DO MATHEMATICAL OBJECTS EXIST?

THE MATRIX

The 1999 science fiction movie *The Matrix* pulls a bait and switch. In *The Matrix*, the hero, Neo, an anagram for the "One," fights an evil organization that has enslaved humanity. Intelligent machines have taken over the world, and they hold people in pods to extract energy from them. To keep the people complacent, the machines insert a simulated reality into the people's brains. The people in the pods do not know that what they see is not real.

Neo escapes and meets Morpheus, the leader of a group of rebels attempting to defeat the matrix and restore people to freedom. Echoing Taoist philosophy, Morpheus asks Neo, "Have you ever had a dream, Neo, that you were so sure was real? What if you weren't able to wake from that dream? How would you know the difference from the real world and the dream world?"[1] We are left to think that Neo is fighting for us, for people who want to live in the real world, but this is not the case. It turns out Neo is just a creation of the intelligent machines, the sixth version of a "Neo" used by the machines to reboot the system. Neo is not real; nothing is real.

The Matrix is fantasy, but Max Tegmark, a professor at the Massachusetts Institute of Technology, takes the idea that reality is a computer simulation seriously. In his 2014 book *Our Mathematical Universe: My Quest for the Ultimate Nature of Reality*, Tegmark argues that the reality we experience is just an imposed mathematical structure. Andrew Liddle writes in his review of the book in *Nature*:

> Tegmark...contends that the Universe is not just well described by mathematics, but, in fact, is mathematics. All possible mathematical structures have a physical existence, and collectively, give a multiverse that subsumes all others.[2]

Such musing may seem far-fetched to the uninitiated but, in fact, this thought has a storied history. The idea that reality is not what it seems goes back to Plato and his Allegory of the Cave. Plato wrote about men locked in a cave, able to see only shadows of outside events on the walls of the cave, never viewing "direct" reality at all. According to this allegory, even if we could determine correctly what is going on outside the cave, we could never be certain of converting our correct opinion into true knowledge.[3]

1 Brandon Toropov and Chad Hansen, *The Complete Idiot's Guide to Taoism* (Alpha, 2002), 241.
2 Andrew Liddle, "Physics: Chasing Universes," review of *Our Mathematical Universe: My Quest for the Ultimate Nature of Reality*, by Max Tegmark, *Nature* 505 (January 2014): 24–25, https://doi.org/10.1038/505024a.
3 Plato, *The Republic Book VII*, trans. W. H. D. Rouse (Penguin, 2015), 365–401.

In Plato's view, the soul lives in one world, the body in another. Physical objects, such as rocks and caves, exist in a world we can touch and feel, but immaterial entities, such as thoughts and feelings, exist in a separate, nonphysical universe. Mathematics is especially prone to Platonistic thinking because mathematical concepts, such as the number "two," cannot be touched. In Platonism, "two" and other mathematical concepts are viewed as "objects" in this nonphysical domain.

The case that "mathematical objects" exist in a world separate from the one we see is presented in *Plato's Problem: An Introduction to Mathematical Platonism* by Marco Panza and Andrea Sereni:

> [We need] to ask an ontological question: are there mathematical objects?... Plato's problem arises in the form of questions such as: do arithmetic and set theory speak of, or describe objects? Are there objects such as natural numbers and sets (i.e., objects that are identical to natural numbers and sets)? Are natural numbers and sets objects?[4]

Panza and Sereni present Plato's argument as a syllogism:

P'.1) Mathematicals exist.

P'.2) Mathematicals are different from physical objects in the sensible world.

P'.3) Mathematicals are ideal objects distinct from sensible things.

Here, Panza and Sereni use the word "mathematicals" to represent all mathematical variables, expressions, equations, and concepts.

---

4   Marco Panza and Andrea Sereni, *Plato's Problem: An Introduction to Mathematical Platonism* (Palgrave Macmillan, 2013), 24.

Can you spot the error in the above syllogism?

Yes, that's right! The major premise assumes that "mathematicals" exist, while we know the primary component of existence is existence itself. Mathematicals are concepts that need to be derived from existence, not pulled out of thin air. To set philosophy right, we must find the relationship between mathematics and things that exist.

"MATHEMATICALS" ARE CONCEPTS

Epistemology is the science devoted to the discovery of the proper methods of acquiring and validating knowledge.[5] To release mathematics from Plato's logical cave, we need to grasp how mathematical concepts are formed.

Step one of the concept formation process was provided in answering *Question 3—What are concepts?* Concepts are formed by focusing the human mind on existents, differentiating a class of existents from all others though their common attributes, omitting the measurements of these common attributes, integrating the result into a unified whole, and labeling the new concept with a word.

With respect to words, Ayn Rand writes:

> The distinguishing characteristic of the new concept is determined by the nature of the objects from which its constituent units are being differentiated, i.e., by their "Conceptual Common Denominator."[6]

What is the Conceptual Common Denominator when forming numbers?

---

5   ITOE, 36.
6   ITOE, 22.

Consider a child sitting at a table on which are placed four groups of objects: a ball, a wooden block, and a doll in one corner, two balls, two wooden blocks, and two dolls in the second corner, three of each in the third corner, and four of each in the fourth corner. An adult may point at the second corner and ask the child, "How many balls are in this corner?" The child may count, "One, two. There are two balls in this corner." Counting is performed by placing each object being considered into a one-to-one correspondence with each number, 1 and 2 using Indo-Arabic numerals, or I and II using Roman numerals, and so on with other number systems.

The Conceptual Common Denominator (CCD) for numbers is the correspondence of the objects in a set of entities with the numerals defined for counting. Numbers differ from words in that it is not "the nature of the objects" that determines the CCD but, rather, the correspondence of the number of objects with the predefined numerals. The CCD for two balls, two blocks, and two dolls is the one-to-one correspondence of the number 2, the second number in the ordered list, with the quantity of objects each set.

Numbers are the "measurements kept" aspect
of the quantity of a particular concept.

The core of arithmetic, indeed the core of all mathematics, is the *Identity Axiom*. Counting from 1 to 2 is represented as:

$$2 = 1 + 1$$

The equal sign states that the left-hand-side (*lhs*) of this equation is *identical* to the right-hand-side (*rhs*). "1 + 1" may look different from "2" in the equation but is, in fact, identical to it. "1 + 1" is just another way to write "2." Since "2" is a number, so is "1 + 1." Keep this in mind when we consider more complicated expressions: No matter the complexity of the expression on the rhs of an equation, it is simply a number if the lhs is a number. That's the Identity Axiom (equal sign) at work.

"Twoness" does not exist in the metaphysical universe. There are two balls, two blocks, and two dolls in the second corner, but no single object called *two*. Further, the two balls (or blocks or dolls) placed on the second corner of the table are not identical. One ball may be red, the other blue; one ball may be large, the other small. Even if the balls look similar at first glance, closer inspection will reveal small variations in one ball compared to the other. The child must form the concept "ball" before he or she learns to count. He or she is counting "concept balls," not actual physical balls. A child must grasp concepts such as ball, block, and doll before he or she can learn the concept "two." Ball, block, and doll are first-level concepts, things you can point at, while "two" is a higher-level concept not existent in perceptual reality.

Higher-level concepts are formed by combining lower-level concepts. Examples of higher-level concepts in the humanities include "justice," "freedom," and "the United States of America." Although words and numbers are derived from things that exist, whether directly or through combinations, neither words nor numbers exist in metaphysical reality. Concepts are tools formed in the human mind to help man organize his knowledge of existents.

Counting began as a tool to keep track of items in existence. For example, early man would keep track of how many sheep entered an enclosure by cutting a slash on a branch for each sheep that passed through a gate:

$$/, //, ///, ////, /////$$

This system is physical: There is a one-to-one correspondence between the number of slashes and the number of sheep. This tally system, or unary numeral system, is cumbersome to use with large numbers. First the Romans, then people in India and Arabia, improved on this notation by the invention of Roman numerals:

I, II, III, IV, V

and of Indo-Arabic numerals:

1, 2, 3, 4, 5

The one-to-one correspondence is conceptual in these systems, as opposed to the physical correspondence employed by early man, but the correspondence is clear: There is a one-to-one correspondence between the number system and the number of existents (such as sheep) that exist in reality. The counted number is incremented by one for every additional existent included in the count. Numbers such as 4,936 and 1.32×10^{15} are impossible to hold as slashes but are uniquely defined using Indo-Arabic numerals and scientific notation.

We define counting as follows:

> *counting*: (v), a conceptual tool employed by men to keep track of the number of existents of a specified type encountered in reality. The tool employs a predefined notation where the count is incremented by 1 with each added existent: *next_number = number + 1*

Counting means order. A number n_1 will be less than another number n_2 provided n_1 precedes n_2 on the list. The inequality

$$n_1 < n_2$$

states that n_1 is located in the counting system earlier than n_2.

Recall Ayn Rand's definition of a concept:

A concept is a mental integration of two or more units possessing the same distinguishing characteristic(s), with their particular measurements omitted, and united by a specific definition.[7]

To form a unit with respect to quantity, man needs to isolate a particular type of entity in his environment. A child needs to isolate the balls on the second corner of the table from the other objects placed there; the farmer needs to isolate the sheep walking into the enclosure from the other animals and objects on the farm. The unit formed is a collection of balls, or sheep, or any other entity, isolated in this way. Mathematicians call a collection of entities in a collection of this type a *set*. Keep in mind that the objects isolated in man's mind are thoughts, not the objects themselves. Every ball is different from every other ball, every sheep is different from every other sheep. It is only man's ability to form concepts that allows him to form the concept "ball," and the concept "sheep," lump the balls and lump the sheep into sets, and count the concepts belonging in each set.

A child will count many sets of existents (actually, he or she is counting the concepts corresponding to the existents): two balls, three blocks, two dolls, six sheep, and many more. The child will notice that the common attribute of a set of two concept balls, a set of two concept blocks, and a set of two concept dolls is two, the second number in the sequence of numbers used for counting; it does not matter what the existents are. The unit for counting is a set of existents possessing a one-to-one correspondence with a predefined counting system.

The counting numbers 1, 2, 3,...are called *natural numbers*. We define the natural numbers as follows:

---

7   ITOE, 10, 13.

A natural number is a mental integration of sets of entities, with the particular measurements of the entities omitted, possessing a one-to-one correspondence in the counting system and labeled by a specific word.

Numbers are neither in the objects being counted, nor preexisting in an alternate reality. Numbers are a mental tool man defines to help him keep track of things in existence. Man can think on the conceptual level where a single word is used to represent an unlimited number of perceptual concretes and a single number is used to represent the quantity of existents.

Panza and Sereni's search for "mathematical objects" is quixotic. Numbers are concepts formed in the mind of man to keep track of the quantity of particular existents perceived conceptually. They are not objects that exist in physical reality.

THE ANSWER

Do mathematical objects exist? No, mathematic objects do not exist. "Threeness" does not appear anywhere, but three objects of the same kind appear many times. Mathematical entities are concepts derived from reality through the concept formation process in man's mind. Once arithmetic and simple geometrical concepts are understood, higher-level relationships are established by using the Identity Axiom.

DOES PLATONISM EXPLAIN ANYTHING IN MATHEMATICS?

HISTORICAL CONTEXT

Consider the mathematical confusion expressed in the Stanford Encyclopedia of Philosophy:

Consider the sentence "3 is prime." This sentence seems to say something about a particular object, namely, the number 3. Just as the sentence "The moon is round" says something about the moon, so too "3 is prime" seems to say something about the number 3. But what *is* the number 3? There are a few different views that one might endorse here, but the platonist view is that 3 is an abstract object. On this view, 3 is a real and objective thing that, like the moon, exists independently of us and our thinking (i.e., it is not just an idea in our heads). But according to Platonism, 3 is different from the moon in that it is not a physical object; it is wholly non-physical, non-mental, and causally inert, and it does not exist in space or time. One might put this metaphorically by saying that on the platonist view, numbers exist "in platonic heaven." But we should not infer from this that according to Platonism, numbers exist in a *place*; they do not, for the concept of a place is a physical, spatial concept. It is more accurate to say that on the platonist view, numbers exist (independently of us and our thoughts) but do not exist in space and time.[1]

The key phrase in this explanation is "3...exists independently of us and our thinking."

The objectivist view is that this is nonsense. The number three does not exist independently of the human mind. Three is a concept formed in man's mind by the process of counting sets of objects containing three existents. Three is a mental tool that helps man keep track of how many existents there are in a set, not an independent quantity in a "Platonic heaven."

The confusion about man's role in defining numbers was exacerbated by Frege in 1884:

The properties which serve to distinguish things from one another are, when we are considering their number, immaterial and beside the point. That is why we want to keep them out of it. But we shall not succeed along the present lines... If, for example, in considering a white cat and a black cat, I disregard the properties which serve to distinguish them, then

---

1   Mark Balaguer, "Platonism in Metaphysics," *The Stanford Encyclopedia of Philosophy*, spring 2016 ed., ed. Edward N. Zalta, https://plato.stanford.edu/archives/spr2016/entries/platonism/.

I get presumably the concept "cat." Even if I proceed to bring them both under this concept and call them, I suppose, units, the whole still remains white just the same... The concept "cat," no doubt, which we have arrived at by abstraction, no longer includes special characteristics of either, but of it, for just this reason, there is only one.[2]

Amazingly, Frege argues that if we extract the "concept cat" from a white cat and from a black cat, we are left with one concept "cat" instead of two. He does not realize that each "concept cat" forms a separate unit, and there are two "concept cats" in the set. Frege concludes that numbers are not to be found in reality, that all you can find is "one" concept cat. Humanity would have been spared a lot of angst had Frege realized the concept "cat" appeared twice in his example, once to represent the white cat, another time to represent the black cat. Although the concept "cat" is unitary, it can be applied any number of times to existent cats.

Having made his mistake, Frege proceeded to lay the groundwork for a new philosophy of mathematics. As described by Rosen:

Frege's insistence that...the truths of mathematics entail that numbers are neither material beings nor ideas in the mind. If numbers were material things (or properties of material things), the laws of arithmetic would have the status of empirical generalizations. If numbers were ideas in the mind, then the same difficulty would arise, as would countless others. (Whose mind contains the number 17? Is there one 17 in your mind and another in mine?)... Frege concludes that numbers are neither external "concrete" things nor mental entities of any sort...they belong to

---

2   Gottlob Frege, *The Foundations of Arithmetic. A Logico-Mathematical Enquiry into the Concept of Number*, trans. J. Austin (Blackwell, Oxford, 1974).

a "third realm" distinct both from the sensible external world and from the internal world of consciousness.[3]

Frege's "third realm" is called *nominalism*, the doctrine that no abstract entities exist. Platonists believe in the existence of abstract objects in a "Platonic heaven"; nominalists argue that no such abstract objects exist. Platonists search for numbers and sets in an abstract reality; nominalists accept only linguistic, symbolic, and logical constructs, very unlike the material world we see. Without abstract entities, and without an understanding of how numbers are formed in man's mind, nominalists are left with symbol manipulation. Nominalism is a subject with no object, as explained in a book by J. P. Burgess and G. Rosen:

> Numbers and other mathematical objects are exceptional in having no locations in space or time or relations of cause and effect. This makes it difficult to account for the possibility of the knowledge of such objects, leading many philosophers to embrace nominalism, the doctrine that there are no such objects.[4]

Even Albert Einstein caught the nominalist bug:

3   José Falguira et al., "Abstract Objects," *The Stanford Encyclopedia of Philosophy*, summer 2022 ed., ed. Edward N. Zalta, https://plato.stanford. edu/archives/sum2022/entries/abstract-objects/. The number 17 is the same in your mind and mine because we all count objects in a set the same way. The number 17 is not in material reality or in "Platonic heaven." If there are 17 items in a set, man can keep track of them by counting them.

4   John P. Burgess and Gideon Rosen, *A Subject with No Object: Strategies for Nominalistic Interpretation of Mathematics* (Clarendon, 1997), 1.

As far as the propositions of mathematics refer to reality, they are not certain; and as far as they are certain, they do not refer to reality.[5]

Prior to Frege, mathematics focused on numbers and the methods used to compute them: analytic geometry, real analysis, calculus, and complex analysis. From the twentieth century onward, mathematicians focused on the structural properties of mathematics: sets, topology, group theory, and symmetry. The leading French mathematician Henri Poincaré wrote:

Mathematicians do not study objects, but the relations between objects; to them it is a matter of indifference if these objects are replaced by others, provided that the relations do not change. Matter does not engage their attention, they are interested in form alone.[6]

The essence of nominalist mathematics is captured in Shapiro's *Philosophy of Mathematics*:

I define a system to be a collection of objects with certain relations... A structure is the abstract form of a system, highlighting the interrelationships among the objects, and ignoring any features of them that do not affect how they relate to other objects in the system.[7]

The nominalist foundation of mathematics is set theory. Provided the mereology is established using sets, the resulting mathematics is

---

5   Albert Einstein, *Ideas and Opinions* (Random House, 1954), 233.
6   Henri Poincaré, *Science and Hypothesis* (Walter Scott, 1905), 20.
7   Shapiro, *Philosophy of Mathematics*, 73–74.

considered purely structural.[8] A structural way to define numbers is through set theory, beginning with the empty set ø, the set that contains nothing. One can then form a progression of sets ø, {ø}, {{ø}}, {{{ø}}},..., where each bracket forms a "set of the set." The sequence of nested sets provide a one-to-one correspondence with the set {0, 1, 2, 3,...}, the counting numbers. In this way, numbers are removed from any reference to reality: Numbers are derived purely from the empty set, a set that does not exist.[9] Any non-repeating progression works for this purpose. The progression ø, {ø}, {ø, {ø}}, {ø, {ø}, {ø, {ø}}},..., proposed by John von Neumann, works just as well.[10] Hellman argues that mathematicians do not need numbers at all.[11] Mathematics, the nominalist argument goes, is an abstract universe created out of nothing.

Note that nominalists are caught in a logical fallacy called a *stolen concept*:

The "stolen concept" fallacy, first identified by Ayn Rand, is the

---

8   Mereology is the philosophical study of the relationships between parts and wholes, or parthood relationships. Achille Varzi, "Mereology," *The Stanford Encyclopedia of Philosophy*, spring 2019 ed., ed. Edward N. Zalta, https://plato.stanford.edu/archives/spr2019/entries/mereology/; and David Lewis, Parts of Classes (Wiley-Blackwell, 1991), 112.

9   E. Zermelo, "Untersuchungen über die Grundlagen der Mengenlhere I," *Mathematische Annalen* 65 (1908): 261–81; see also Jean Van Heijenoort, ed., *From Frege to Gödel: A Source Book in Mathematical Logic, 1879–1931* (Harvard University Press, 1967), 199–215.

10   John von Neumann, "Zur Einführung der transfiniten Zahlen," *Acta Litterarum ac Scientiarum: Sectio Scientiarum Mathematicarum* 1 (1923): 199–208; and English translation "On the Introduction of Transfinite Numbers," in *From Frege to Gödel: A Source Book in Mathematical Logic, 1879–1931*, ed. Jean Van Heijenoort (Harvard University Press, 1967), 346–54.

11   Geoffrey Hellman, *Mathematics Without Numbers: Towards a Modal-Structural Interpretation* (Oxford University Press, 1989), 48–49.

fallacy of using a concept while denying the validity of its genetic roots, i.e., of an earlier concept(s) on which it logically depends.[12]

Sets are formed in the mind of man *by reference to things that man experiences in reality*. The empty set is formed by considering two identical non-empty sets A and B and asking the question: What happens if we subtract A from B? The concept "empty set" could not be formed if the concept "non-empty set" had not been formed.

MATHEMATICS AS A FICTION

We will explore the nature of higher-level mathematical concepts in answering the Ultimate Questions to follow. For now, from a historical perspective, we note that nominalism led to *fictionalism*. Hartry Field held that since abstract objects do not exist, mathematical statements are not true, or are only vacuously true.[13] Mathematical statements are true only in the same sense that ordinary language is said to be true. We create a convenient fiction, this line of reasoning holds, where mathematical truths are true in the same way as acts by a fictional character are true. For example, the statement "Howard Roark blew up Cortlandt Estates" is a true statement, but it occurs in the novel *The Fountainhead*. Field argues we can speak of truth about mathematical statements provided we understand this truth is only about a story mankind made up about the world.[14]

An inconvenient truth about all these theories is that mathematics works. Nominalism and fictionalism may be self-consistent, but they fail to explain why mathematics has the power to explain the quantities we measure. The pragmatist says, *bunk to your theories,*

---

12 PWNI, 22.
13 Hartry Field, *Science Without Numbers* (Blackwell, 1982).
14 Hartry Field, *Realism, Mathematics and Modality* (Blackwell, 1989).

mathematics works, and that's all that matters. This way of thinking leads to the Indispensability Argument, the idea that mathematics is real because it provides real results. As sketched by Panza and Sereni, the Indispensability Argument is:

> Suppose that there are true scientific theories, or at least scientific theories that we are justified in holding true... Since it is natural to conclude that the theories in question could not be true if these mathematical objects did not exist, it follows that these objects exist, or that we are justified in holding that they exist.[15]

Panza and Sereni hold that Platonism is true; at least it is true for mathematics. They employ spectacularly convoluted pretzel logic to justify their belief:

> The role of a criterion of ontological commitment within IA [the Indispensability Argument] for Platonism depends on the natural assumption that a theory T is true, or we are justified in believing it true, only if there actually exists, or we are justified in believing that there actually exists, everything that T says exists, from which it follows that T is true, or we are justified in believing that T is true, only if there actually exists, or we are justified in believing that there actually exists, any object included in T's ontological commitment.[16]

In other words, Panza and Sereni have no idea why mathematics works, but, since it does work, they claim that "mathematical Platonic heaven" must exist. This is akin to early man not understanding why

---

15   Panza and Sereni, *Plato's Problem*, 197.
16   Panza and Sereni, *Plato's Problem*, 211.

it rains on some days and not others and hypothesizing a "rain god" to explain the rain.

The classical alternative to Platonistic idealism is Aristotelian realism. Aristotelian realism has been largely ignored by mathematicians over the past few centuries but has found strong support in James Franklin's *An Aristotelian Realist Philosophy of Mathematics*. Franklin confronts the difficulty Aristotelians face:

> The challenge for the Aristotelian philosopher of mathematics is...to explain how pure mathematics is really about universals that could be and sometimes are realized in (non-abstract, possibly physical) reality.[17]

However, Franklin concedes Aristotelians fail to meet the challenge:

> The thesis defended has been that some necessary mathematical statements refer directly to reality. The stronger thesis that all mathematical truths refer to reality seems too strong... The Aristotelian can admit that negative numbers, the square root of minus 1, the average Londoner and other such entities could be fictions.[18]

The Aristotelian view that mathematical universals are realized in "non-abstract, possibly physical" reality, is untenable. Ideas do not exist in objects. Further, there is no Platonic heaven. The only logical possibility is the objective one: Mathematical concepts are formed in man's mind by observing the relationships between quantities in existence.

---

17   James Franklin, *An Aristotelian Realist Philosophy of Mathematics: Mathematics as the Science of Quantity and Structure* (Palgrave Macmillan, 2014), 51.

18   Franklin, *An Aristotelian Realist Philosophy of Mathematics*, 81.

ARITHMETIC

Mathematics is susceptible to Platonistic thinking because all mathematical concepts are derived from abstractions. Even forming the number "2" requires a process of thought, extracting the concept "quantity two" from multiple instances of two entities under the same concept in perceptual reality. To put mathematics on a solid footing, we need to examine the logical chain by which high-level mathematical concepts are formed.

Leonard Peikoff notes:

> ...concept formation proceeds by a process of abstracting from abstraction. The result is (increasingly) higher-level concepts, which cannot be formed directly from perceptual data, but only from earlier concepts...Concepts, therefore differ from one another not only in their referents, but also in their distance from the perceptual level.[19]

Although the distance of high-level mathematical concepts from the perceptual level is large, the logical chain by which these concepts are formed must be intact.

Let's begin with arithmetic. If you walk 3 steps east and then walk a further 5 steps east, you have walked 8 steps east. This leads to addition:

$$c = a + b$$

where a, b, and c are numbers. If you walk 3 steps east and then walk 5 steps west, you have walked −2 steps west. Note that negative two is realized physically in this example; it is simply stepping west with east as the positive direction. Subtraction is the inverse of addition:

---

19   OPAR, 91.

$$a = c - b$$

Now consider walking 3 steps east and then walking 3 steps west. You are back to where you started: The net distance you walked is 0 steps even though you took a total of 6 steps. Zero is just like any other number on this scale. You can point to the zero-step location on the ground just as easily as you can point to the one-step location.

The ancient Egyptians had a symbol for zero, but the ancient Greeks did not. The Greeks did not believe that nothing could be something. They did not realize that counting is a mental construction used to keep track of things, that zero is a number like the others. You will never be able to point to zero items in existence, but if you take three steps forward and three steps back, you are zero steps away from your starting point.

No number, not 1, 2, 3, or 0, refers to an existent in metaphysical reality. Numbers are conceptual tools used to keep track of the number of things you conceptualize. The validity of a number depends on two things:

1. Does the number provide a useful tool for keeping track of entities in existence?
2. Does the number satisfy the Identity Axiom?

The number 2 is valid because there is a one-to-one match between the number and the number of items counted. The number 0 is valid because the equation:

$$c = b - b = 0$$

satisfies the Identity Axiom.

Once man developed addition, multiplication and division follow:

$$c = a \bullet b \qquad a = c / b$$

Multiplication is just repeated addition; division, repeated subtraction.

Arithmetic and other mathematical concepts are "concepts of method" as defined by Ayn Rand:

> Concepts of method designate systematic courses of action devised by men for the purpose of achieving certain goals...Concepts of this category have no direct referents on the perceptual level of awareness.[20]

Concept methods must satisfy the following rule:

To be valid, a concept of method must obey the Identity Axiom.

The Identity Axiom ties concept formation and existence together. A method that violates the Identity Axiom will give results that differ from observed reality. With sufficient inspection, a mathematical concept that obeys the Identity Axiom can be traced back to referents in existence. The world of "method concepts" is not Platonic. Correct mathematical methods have a valid relationship with things that exist. Mathematics provides a unified structure:

> No concept man forms is valid unless he integrates it without contradiction into the total sum of his knowledge.[21]

The essential point is this: Contradictions do not exist. No part of mathematics can contradict any other part.

---

20   ITOE, 35–36.
21   FNI, 126.

THE ANSWER

Does Platonism explain anything in mathematics? No, it expresses profound ignorance. Numbers and all of mathematics have a clear, defined relationship with reality.

ARE THE SQUARE ROOTS OF 2 AND OF −1 NUMBERS?

THE INDIANA PI BILL

On February 6, 1897, the Indiana House of Representatives passed a bill declaring that in Indiana, the value of π will henceforth be equal to 3.2. Now known as the Indiana Pi Bill, sanity was saved by Professor C. A. Waldo of Purdue University who arrived in time to stop the Indiana Senate from passing the bill. The law never actually went into effect.[1]

The law also declared the value of √2 to be equal to 10/7 = 1.429.

What is wrong with passing such a law? Yes, the approximations π and of √2 are crude. One can compute more digits to π and √2, say π = 3.141926 and √2 = 1.414214, and pass a better law. Others say you

---

1   Wikipedia, "Indiana Pi Bill," accessed December 6, 2024, https://en.wikipedia.org/wiki/Indiana_Pi_Bill.

need even more digits, say π = 3.14159265358979932384626433 and √2 = 1.41421356237309950488016887, surely that's enough! No man will ever require more accuracy than that!

The problem with the Indiana Pi Bill is epistemological. It places man's thoughts, rather than reality, in the driver's seat. The Indiana Pi Bill was ridiculed in 1897, and it is ridiculous today. Ideas should be derived from reality, not the other way around. The idea that π or √2 can be represented by any number of decimal digits is a primacy of consciousness approach to mathematics. The primacy of existence approach says that π, √2, and all other irrational numbers cannot be represented by decimal digits no matter how many you use.

According to Leonard Peikoff:

A concept denotes facts—as processed by a human method. [It does not] introduce any cognitive distortion. The concept does not omit or alter any characteristic of its referents.[2]

The Indiana Pi Bill distorted the concept of π, but so does the common idea that π is a sequence of digits on the number line. The length of the sequence is irrelevant; even a million decimal-place expansion of π does not give π as the ratio of the circumference of a circle to its diameter. No matter how many digits you use, the expression you write will not equal π.

Definitions are a crucial step in concept formation. As Ayn Rand explains, the basic function of a definition is:

...to distinguish a concept from all the other concepts and thus keep its units differentiated from all other existents.[3]

---

2   OPAR, 89.
3   ITOE, 28.

These differentiae are not arbitrary. According to Peikoff:

> ...we can differentiate only on the basis of a wider characteristic, the CCD, which is shared both by the concretes we are isolating and by the concretes from which we are isolating them.[4]

The Conceptual Common Denominator for irrational numbers is that they cannot be written as a series of digits. To try to write them as a series of digits destroys the CCD.

The square root of 2 is the number that multiplied by itself gives 2. If we multiply 1.4 by itself, we get 1.4 × 1.4 = 1.96; if we multiply 1.5 by itself, we get 1.5 × 1.5 = 2.25. We may not know the value of the square root of 2, usually written as √2, but its value must be between 1.4 and 1.5.

From the time of the ancient Greeks, mathematicians have known that square root 2 cannot be expressed as a rational number, where a rational number is any number expressed as a ratio of integers. A geometric proof is given in Euclid's *Elements*. A more straightforward proof involves algebra. Assume that √2 can be written as a/b where both a and b are integers and reduce a/b so it has no common factors. Next assume a is an even number and, after a few steps, conclude if a is even, then b must be even also. However, if both a and b are even, the ratio a/b has a common factor, namely 2, and we have a contradiction. Contradictions cannot exist; thus, it is not possible to write √2 as a/b.

Numbers such as √2 that cannot be written in the rational form a/b are called *irrational numbers*. Since square root 2 cannot be written as a rational number, its decimal expansion goes on forever without repeating. But forever does not exist in reality; no one will ever write down the decimal expansion √2 to complete accuracy. This leads to the claim, widely circulated on the internet, that √2 is not a number, that

---

4   OPAR, 97.

one can approximate square root of two to any desired degree of accuracy but that the exact value of √2 is impossible to find. In a twist of logic, "impossible to find" turns into "does not exist." The idea "nothing can exist that is unavailable to man's mind"—an idea that makes existence subservient to consciousness—leads to: *If I cannot write down the exact value of √2 using decimal arithmetic, √2 does not exist.*

Arguing that the square root of two is not a number is like legislating the value of pi. It destroys the concept. Only a primacy of consciousness worldview—a worldview that places consciousness above existence—allows preconceived notions such as "all numbers must be written as a series of digits" to rise over reality. The primacy of existence worldview requires that we accept facts as they are.

Robert Knapp discusses this point extensively in *Mathematics Is About the World: How Ayn Rand's Theory of Concepts Unlocks the False Alternatives Between Plato's Mathematical Universe and Hilbert's Game of Symbols.* To begin, he notes that approximation and measurement are inexorably linked. Saying, "I have an approximate value of √2 as 1.412" and saying, "I have measured the value of √2 as 1.412," are essentially equivalent: Both statements say, "√2 exists; I have a close but imprecise value for it." Knapp writes:

> Numbers designate specific mathematical relationships; any two different numbers potentially identify distinguishable quantitative relationships... Any rival approximation, different from √2, would, at some point, fall outside the precision range... In the unqualified sense of number, there is no range [of approximate values] because, as far as the concept of number is concerned, the precision standard is an omitted measurement. Every number, as such, names a distinct mathematical relationship...distinguishable from any other number.[5]

---

5   Robert E. Knapp, *Mathematics Is About the World: How Ayn Rand's Theory...*

The exact value of √2 written in decimal notation to full precision does not exist. Only one number satisfies the equation:

$$a \bullet a = 2$$

No number except √2 gives 2 when it is multiplied by itself; no approximation of √2 gives 2 when multiplied by itself. The number 2 is a concept derived in man's mind by observing existents. The number √2 is a concept derived in man's mind through the Identity Axiom. Robert Knapp writes:

> The difference between rational numbers and irrational numbers does not represent a metaphysical distinction among *magnitudes*: the difference between rational and irrational numbers consists in the means of measurement, in how such numbers are specified, not with the object of measurement... Qua number, precision is an omitted measurement; both rational numbers and irrational numbers designate specific, distinguishable, mathematical relationships.[6]

The natural numbers are "concepts formed directly by interacting with existence through perception." Irrational numbers are "concepts of method." Although irrational numbers differ from the natural numbers in that they are not available directly by counting, both are useful and valid concepts.

Man's mind forms two types of numbers to measure and model reality: rational numbers and irrational numbers.

...of Concepts Unlocks the False Alternatives Between Plato's Mathematical Universe and Hilbert's Game of Symbols (CreateSpace Independent Publishing, 2014), 238, 243.

6   Knapp, *Mathematics Is About the World*, 238.

A similar issue arises with the square root of negative 1:

$$a \bullet a = -1$$

No real number satisfies this equation. However, if we let $a = i$ be a mathematical entity where $i = \sqrt{-1}$, then we obtain a very useful "method concept," because i is ubiquitous in modeling physical systems.

Ayn Rand addressed the validity of mathematical concepts such as $\sqrt{-1}$ directly in response to Professor C in the discussion following her publication of *Introduction to Objectivist Epistemology*. Professor C says that imaginary numbers are useful in computations but:

> I personally do not see the validity of this concept. There is nothing in reality to which it corresponds. Nothing is measured except by real numbers.[7]

After a short discussion, Ayn Rand answers:

> Whenever in doubt...about the standing of any concept, you can do what I have done in this discussion right now. I asked you, "What, in reality, does that concept refer to?" If you tell me that the concept, let's say, of an imaginary number doesn't do anything in reality, but somebody builds a theory on it, then I would say it is an invalid concept. But if you tell me, yes, this particular concept, although it doesn't correspond to anything real, does achieve certain ends in computations, then clearly you can classify it: it is a concept of method, and it acquires meaning only in the context of a certain process of computation.[8]

---

7   ITOE, 305.
8   ITOE, 306.

Concepts of method, such as the number √2 or the number *i*, have no direct referents in perceptual reality. This does not make them Platonic. *Irrational numbers and imaginary numbers are concept of method numbers with unique, well-defined relationships to the natural numbers.* The chain linking irrational numbers and imaginary numbers to things that exist is conceptual but unbroken. No number, not even the natural numbers, are given metaphysically. Real numbers and imaginary numbers have the same ontological status: all numbers, including the natural numbers, are higher-level concepts formed in the mind of man to keep track of things that exist.

THE ANSWER

Are the square roots of 2 and of −1 numbers? Yes, the square roots of 2 and of −1 are numbers. They cannot be written in decimal form, but they have unique, well-defined relationships with the counting numbers 1 and 2. Their existence is a result of the Identity Axiom. Ontologically, their status is the same as that of the counting numbers. Neither the square roots of 2 and of −1 nor the counting numbers exist as objects in the universe. All numbers are abstract concepts formed in man's mind to assist in understanding reality.

IS INFINITY A NUMBER?

INFINITY

nfinity does not exist in perceptual reality. You will never write down an infinite string of numbers, count an infinite number of stars, or perceive an infinite number of seconds. There can be no last number; no matter how large the number you find, you can always add 1 to it. Numbers go on forever. As Robert Knapp notes:

> That one can reach any multitude by counting high enough means that there is no actual infinity. All multiplicity is finite. An infinite number would have to be one that could not be reached by counting.[1]

The puzzlement with infinity was recognized by Zeno of Elea in the fifth century BCE. In one of Zeno's paradoxes, a man attempts

---

1   Knapp, *Mathematics Is About the World*, 230.

to walk across a room but is unable to get to the other side. Zeno said that to walk across the room, the man must first get halfway there. Once he is halfway across, he must traverse halfway across the remainder, namely ½ • ½ = ¼ of the way across the room. Next, he must walk across half of the remainder, namely ½ • ¼ = ⅛, and so on, over and over. Each time the man reaches the next halfway point, he has half of the remaining distance to go.

The distance the man travels after n iterations of this process is to add the "halfway" distances across the room, according to the sum:

$$d_n = ½ + 1/2^2 + 1/2^3 + 1/2^4 + \dots + 1/2^n = \textit{Sum of}[\,1/2^i\,]\ \textit{for}\ i = 1\ \textit{to}\ n$$

In the computer era, this sum is computed using an algorithm:

```
distance = 0;
step = 1/2;
For [n = 0, distance < 1, n + +,
  distance = distance + step;
  step = 1/2 step;
Print[distance];
  ];
```

A Mathematica computer program that computes the Zeno series and the value of the "distance travelled across the room" for fifteen steps is presented in Figure 1. We see that the sequence of numbers is converging toward one but never reaches it; "distance" is less than one no matter how large we make n. Everyday experience says that the man crosses the room; Zeno's paradox says he can never get there.

The paradox is resolved by understanding the difference between the concept of infinity as a one-to-one correspondence between counting existents, and infinity as a concept formed through repetition. Consider the number 1/3. While 1 and 3 are natural numbers, 1/3 is not. You are not able to cut an apple into exactly three pieces.

```
distance = Table[ Sum[ ( 1 / 2 )^n, { n, i}], {i, 15} ];
distance = Flatten[ Append[ { 0 }, distance ] ];
Print[ "Distance across room = ", N[ distance ] ];
ListLinePlot[ distance, AxesOrigin → {1, 0 } ]
```

```
Distance across room = {0., 0.5, 0.75, 0.875, 0.9375, 0.96875, 0.984375,
   0.992188, 0.996094, 0.998047, 0.999023, 0.999512, 0.999756, 0.999878, 0.999939, 0.999969}
```

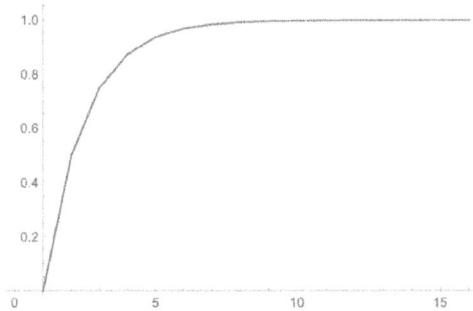

Figure 1. A Mathematica computer program that computes distance traveled
by the man in the Zeno paradox, as well as the calculated and graphed results.
Note that the distance traveled approaches 1.0 but never reaches it.

Even if you have three apples, because each particular apple is different, you will never be able to distribute the three apples in an exactly
equivalent way.

When I say, "I have three apples; here is one-third of them," I mean,
"I have three items in existence that I identify as apples; I am giving
you one-third of the existents I identify under the concept apple."

The "concept apples" are distributed exactly in thirds; the
physical apples are distributed only approximately in thirds.

Keep in mind that words and numbers are only tools to help us
deal with existence. One-third is a "method concept" that is very useful to man but is not found in metaphysical reality.

To write 1/3 as a decimal expansion, we first divide 1.0 by 3 to
obtain the quotient 0.3 and the remainder 0.1. Dividing the remainder

0.1 by 3 gives the new quotient 0.03 with the new remainder 0.01. At this point, we have $1/3 = 0.3 + 0.03 + 0.01 = 0.33 + 0.01$. Repeating the process n times gives:

$$1/3 = 0.333333\ldots + 1.0 \times 10^{-n}$$

where the length of the string of 3s is n. There is no end to this process; it goes on forever. In practice, we cannot write a string of 3s that goes on forever; but on the conceptual level, there is no reason to stop. On the conceptual level, 1/3 and 0.333333...(where the 3s go on forever) are identical:

$$1/3 = 0.333333\ldots$$

The Identity Axiom requires that we form the concept "infinite string of 3s" to express 1/3 in decimal notation. Note that this is an aspect of base 10 arithmetic; in base 3 arithmetic, 1/3 is written as 0.1. Infinity in a mathematical context is a "method concept" required by the Identity Axiom. It does not imply that an infinite string of 3s exists in metaphysical reality. Of course, neither does 1/3.

Ayn Rand's observation that concepts are open-ended is relevant here:

An arithmetical sequence extends into infinity, without implying that infinity actually exists; such extension means only that whatever number of units does exist, it is to be included in the same sequence. The same principle applies to concepts: the concept "man" does not (and need not) specify what number of men will ultimately have existed—it specifies only the characteristics of man, and means that any number of entities possessing these characteristics is to be identified as "men."[2]

2   ITOE, 18.

The concept "repeated 3s" is open-ended; the number of repeated 3s is not part of the concept. The use of three dots (...) to indicate a continuing sequence of numbers indicates repetition.

Augustin-Louis Cauchy resolved Zeno's paradox in 1827.[3] In Cauchy's method, the convergence of a sequence of numbers is established by considering the difference in magnitude between the number d_n at the nth step of the sequence and the value it converges toward but never reaches. As an example, let d_n be the sequence of numbers in the Zeno paradox. From Figure 1, we see this sequence approaches 1 as *n* gets large, so we use 1 as the expected convergence value. The magnitude mag_n of the difference between d_n and 1 is:

$$mag_n = |\, d_n - 1\, |$$

where the vertical brackets mean that mag_n is always positive; i.e., if $d_n - 1$ is negative, change the sign so that mag_n becomes positive. It is possible to show algebraically that mag_n gets smaller as *n* grows. Therefore:

mag_m is less than mag_n for any m greater than n

Now let's choose a small positive number, usually denoted by the Greek letter *epsilon*, to define the precision we desire in the computation. As an example, let's choose *epsilon* = 0.001 in the Zeno paradox sequence. We then increase the value of *n* until mag_n is less than *epsilon*. And since we know that mag_m is less that mag_n for any value of *m* greater than *n*, it follows that every number in the sequence mag_m will be smaller than *epsilon* regardless of how large we make *m*.

---

3   Augustin-Louis Cauchy, "Sur un nouveau genre de calcul analogue au calcul infinitesimal," *Exercices de mathematiques*, vol. Seconde Année. (1827); repr. Cambridge University Press (2009), https://doi.org/10.1017/CBO9780511702655.004.

The above argument works regardless of how small you make *epsilon*. In the limit as *epsilon* approaches zero, the Zeno sequence d_n converges to 1.

The Cauchy method never actually uses infinity in the computation; *m* and *n* are always finite. The sequence d_n never reaches 1 for any finite value of *n*. One may argue: "After 10 steps, the remaining distance is $(\frac{1}{2})^{10} = 0.00098$. This is so small; d_n for $n = 10$ equals 1 for all practical purposes. What difference does it make if I set d_{10} equal to 1?"

The difference is adherence to the Identity Axiom. Concepts of method that violate the Identity Axiom throw us into the neverland of imprecision. Setting d_{10} equal to 1, or d_{100} equal to 1, or even $d_{1,000,000}$ equal to 1, violates the Identity Axiom. Only $d_\infty = 1$, where ∞ is larger than any finite number, satisfies the Identity Axiom. We must extend the number of "halfway" intervals across the room to infinity to obtain a result consistent with the Identity Axiom. If you get rid of the Identity Axiom, anything goes.

From an engineering standpoint, we would say, it is close enough: $(2^n - 1)/2^n = 1$ for large enough *n*. Notice, however, that to achieve this result, we set $2^n - 1 = 2^n$. Thus, for a large enough *n*, the engineering approximation grants 2^n the mathematical property of infinity:

$$\textit{infinity} + \textit{any\_finite\_number} = \textit{infinity}$$

Infinity is the concept formed by neglecting the addition or subtraction of any finite quantity. By this definition, infinity cannot be a number. The difference between a number and its successor must equal one.

The numerical value of $\sqrt{2}$ can also be evaluated by means of an infinite series. In this case, an infinite series that converges to $\sqrt{2}$ is:[4]

$$\sqrt{2} = 1 + \tfrac{1}{2} - 1/(2 \bullet 4) + (1 \bullet 3)/(2 \bullet 4 \bullet 6) - (1 \bullet 3 \bullet 5)/(2 \bullet 4 \bullet 6 \bullet 8) + \ldots$$

---

4   This polynomial series is a relatively inefficient method for computing $\sqrt{2}$.

The first twenty values of this sum are computed by the following algorithm:

```
sqrt_2 = 1;
numerator = 1;
denominator = 2;
sign = -1;
For [ n = 1, n < 20, n++,sign = -sign;
  increment = sign numerator / denominator;
  sqrt_2 = sqrt_2 + increment;
  numerator = (2n - 1) numerator;
  denominator = 2 (n + 1) denominator;
  ];
Print[ sqrt_2];
```

The sequence of numbers produced by this algorithm is plotted in Figure 2. Note that the values in this series oscillate above and below the value √2. Square root of two is bracketed by the upper and lower bounds as accurately as desired. However, the limit of this Cauchy sequence, namely √2, is never reached. To make mathematics complete, we must grant ontological status to the limits of Cauchy sequences.

Robert Knapp observes:

...the use of infinite series provides a general mathematical approach to successive approximation. But, because Cauchy sequences can converge to irrational numbers, one cannot develop a theory of approximation that doesn't include irrational numbers as potential limits of a sequence of approximations... The validity and universality of mathematical conclusions depends on the ability to analyze complex chains of mathematical relationships without ever losing precision.[5]

5   Knapp, *Mathematics Is About the World*, 252–53.

$\sqrt{2}$ => {1.5, 1.375, 1.4375, 1.39844, 1.42578, 1.40527,
  1.42139, 1.40829, 1.4192, 1.40993, 1.41794, 1.41093, 1.41713,
  1.4116, 1.41658, 1.41206, 1.41618, 1.41241, 1.41588, 1.41267}

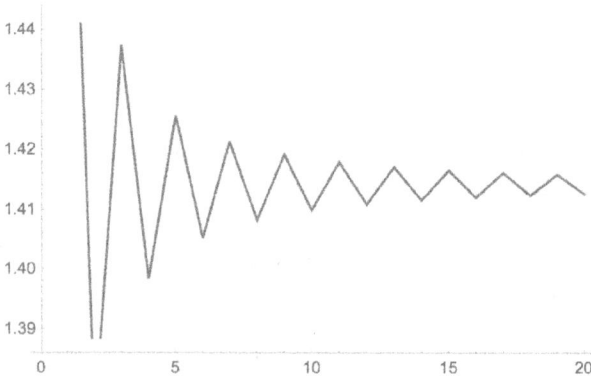

Figure 2. The first twenty values of the polynomial series expansion of $\sqrt{2}$ and a corresponding plot. Note that the numerical values oscillate above and below the correct value of $\sqrt{2}$ which is computed to six decimal places.

Square root of two is the unique number that satisfies the equation $a \cdot a = 2$. It is also the limit of the above Cauchy sequence.

THE ANSWER

Is infinity a number? No, infinity is not a number. Infinity is a "measurement omitted" concept that specifies that a process is to be repeated many times without specifying how many times it is to be repeated. The limits of Cauchy sequences are granted ontological status to make mathematics complete.

IS ARITHMETIC AXIOMATIC?

THE PLATONIST APPROACH TO MATHEMATICS

n Platonism, numbers exist independently of the human mind. The Platonist does not arrive at 1, 2, 3 by counting. He assumes numbers are preformed in a Platonic heaven and, like the captives in the allegorical cave, he sees them using his intuition. In the modern era, such "reality-free mathematics" is commonly generated by defining an axiomatic structure for mathematics, either directly or by using a set theory.

In 1889, Giuseppe Peano published a set of axioms used by modern mathematicians to define arithmetic.[1] These axioms are:

---

1   Michael Segre, "Peano's Axioms in Their Historical Context," *Archive for History of Exact Sciences* 48, no. 3 (September 1994): 201–342, https://doi. org/10.1007/bf00375085.

1. Zero is a natural number.
2. For every natural number x, $x = x$.
3. For all natural numbers x and y, if $x = y$, then $y = x$.
4. For all natural numbers x, y, and z, if $x = y$ and $y = z$, then $x = z$.
5. For all a and b, if b is a natural number and $a = b$, then a is also a natural number.
6. For every natural number n, $S(n)$ is a natural number.
7. For all natural numbers m and n, $m = n$ if and only if $S(m) = S(n)$.
8. For every natural number n, $S(n) = 0$ is false.
9. If K is a set such that:
    - 0 is in K, and
    - for every natural number n, n being in K implies that $S(n)$ is in K, then K contains every natural number.

Here, $S(n)$ is the successor function:

$$S(n) = n + 1$$

As described in answering *Question 10—Do mathematical objects exist?*, numbers are derived from reality by counting things. For example, the number 3 provides a one-to-one correspondence between the counting system and 3 existents identified under the same concept. Basic aspects of reality are that $3 - 3 = 0$ (Peano axiom 1), $3 = 3$ (Peano axioms 2 and 3), $3 + 1 = 4$ (Peano axiom 6), and $3 = 3$ implies $4 = 4$ (Peano axiom 7). Axioms 4 and 5 are trivial when written in terms of numbers. Axioms 1 and 8 together start the count at 0, an unnatural choice for counting but convenient for other purposes. Axiom 9 states that the set containing all natural numbers exists.

Thus, aside from two quirks, beginning the natural numbers with 0 instead of 1, and assuming the existence of the infinite set K, *the Peano axioms are properties of the natural numbers, not their definition*. This is analogous to defining a musical melody using words, saying it is harmonious and soothing, rather than actually playing the notes. Peano created a structure without a reference to the concepts that gave rise to the structure in the first place.

The importance of connecting concepts to their origins in existence cannot be overstated. In Ayn Rand's words:

> To know the exact meaning of the concepts one is using, one must know their correct definitions, one must be able to retrace the specific (logical, not chronological) steps by which they were formed, and one must be able to demonstrate their connection to their base in perceptual reality.
>
> When in doubt about the meaning or the definition of a concept, the best method of clarification is to look for its referents—i.e., to ask oneself: What fact or facts of reality gave rise to this concept? What distinguishes it from all other concepts? [2]

Axiomizing arithmetic is impossible without examining how people use numbers. Peano may not have known the connection between numbers and the facts of reality that give rise to numbers, but he knew how numbers worked. Arithmetic came first, the axioms afterward. Peano reverse engineered the known properties of numbers to obtain his axioms. To claim that mathematics is derived from the Peano axioms is an example of the stolen concept fallacy. The Peano axioms are a derivative, not a primary, approach to defining numbers.

Existence, entities, and consciousness are axiomatic. Numbers are derivative. As described in answering *Question 10*, numbers are a consequence of man's mind looking at entities in existence and recognizing multiple instances of entities labeled under the same

---

2   "What fact or facts of reality gave rise to this concept?" is commonly called *Rand's question*. ITOE, 51.

concept. Peano's "axioms 1 through 8" provide correct properties of numbers, but *these properties are derived from observation of reality and, therefore, are not axiomatic.* Peano's axiom 9 is really about set theory. We examine set theory next.

Ernst Zermelo and Abraham Fraenkel formalized set theory in the early twentieth century with a set of axioms. The ZF axioms, as they are known today, provide a structure of interlocking sets, partitioning the universe into groups of any form or content, and define the rules for unions and intersections of sets. Set theory creates a new language of symbols through which theorems and proofs are built. This edifice turns mathematics into pure thought of no value to anyone outside the discipline. If sets describe a closed, complete system, then there is no need for a mathematician to interact with the outside world.

For all the complexity imposed by the ZF axioms, a set is just counting in groups. Keeping track of things in groups has been done for thousands of years. During the Imperial period, a Roman legion was composed of ten cohorts, each cohort comprised of six centuries, each century having one hundred soldiers. When a Roman general united his set of five legions with five new legions sent by Rome, he had ten legions providing sixty thousand troops. When he discovered that Rome had mistakenly included one of his legions in the assignment of new legions, he calculated the intersection of his existing legions and his newly assigned legions, and concluded it was one. He was six thousand soldiers short.

One can think of a million such examples. The idea of grouping objects into sets is trivial. Yet modern mathematics places set theory at center stage. How did this trivial concept, the concept that entities can be organized into groups and groups compared, combined, and split, take over mathematics?

As described in answering *Question 11—Does Platonism explain anything in mathematics?*, set theory attempts to create mathematics out of nothing. Placing the null set and the rules for combining sets as the foundation of mathematics severs mathematics from the world. As Robert Knapp notes:

> From a reality-based perspective this is simply crazy. It completely cuts off mathematics from any official relationship to the world. It ignores the context of how sets actually arise in mathematics, why they are needed, what they mean, and how is it that they actually provide the distinctions that they are designed to provide. To adopt the ZF axioms as a foundation of mathematics is to abandon, on principle, any substantive content of mathematics.[3]

In effect, mathematicians steal the concept of a set from everyday experience and then trivialize it by claiming the null set ø is the foundation of mathematics. The whole process is unreal:

> ...the entire machinery of set theory is taken, in a sense, with a grain of salt. Mathematicians treat numbers as if they had nothing to do with the weird constructions from the ZF axioms. Yet they take comfort in the formal equivalence of these constructions with the numbers that everyone uses. For this formal equivalence seems to show that one can go on using numbers, ordinary logic, and the prescribed operations on sets, without fear of contradiction. In effect, they treat set theory as a useful model, not as the context of their enterprise.[4]

As is the case with the Peano axioms, the ZF axioms do not qualify as axioms. An axiom is "...a primary fact of reality, which cannot be analyzed, i.e., reduced to other facts or broken into component

---

3   Knapp, *Mathematics Is About the World*, 350.
4   Knapp, *Mathematics Is About the World*, 351.

parts."[5] The ZF "axioms" specify *properties* of sets, easily derived by forming groups of entities rather than counting them individually. A child forms the concept "number" by placing concepts representing similar entities into a group or "set" and setting up a one-to-one correspondence with a counting system.

The concept "set" is derived from reality, not the other way around.

OBJECTIVE MATHEMATICS

Bertrand Russell famously said:

> Mathematics may be defined as the subject where we never know what we are talking about, nor whether what we are saying is true.[6]

The ignorance expressed in this statement is extreme. Yet Russell is viewed as one of the architects of modern mathematical thought.

Russell's error, and that of mathematicians generally, is to begin with mathematical axioms instead of philosophic axioms. Philosophic axioms are the base of all knowledge; mathematical axioms are not really axioms, but properties derived from existence. Since existence is primary, and consciousness is awareness of existence, mathematics must conform to man's awareness of existence; mathematics cannot be created by fiat.

Like many others, I studied Euclidian geometry in high school and learned the five Euclidian axioms. These axioms seemed sensible to me at the time. Yes, I thought, that's how space works. Euclid's five axioms are:

5   ITOE, 55.
6   Bertrand Russell, *Mysticism and Logic* (Allen & Unwin, 1917), 75.

1. It is possible to draw a straight line from any point to any point.
2. It is possible to extend a finite straight line continuously in a straight line.
3. It is possible to describe a circle with any center and radius.
4. All right angles are equal to one another.
5. If a straight line falls across two straight lines and the interior angle on the same side is less than two right angles, the two straight lines, if extended indefinitely, will meet on that side on which the angles are less than two right angles.[7]

But these axioms are not enough. Euclidian proofs also require the following five properties:

1. Things that are equal to the same thing are also equal.
2. If equals are added to equals, then the wholes are equal.
3. If equals are subtracted from equals, then the differences are equal.
4. Things that coincide with one another are equal.
5. The whole is greater than a part.[8]

Is geometry axiomatic? Or are these ten truths properties of existence derived by thinking about space?

Euclid's first "axiom," It is possible to draw a straight line from any point to any point, is a concept derived from existence such as using a stretched rope to measure the distance between two pegs. Similar arguments hold regarding axioms 2, 3, and 4 and the five properties listed above. We know that axiom 5, the parallel postulate, applies

---

7   Euclid, *Elements*, book I, proposition 5, trans. Thomas Little Heath (University Press, 1908), 195–202.

8   Gerard A. Venema, *Foundations of Geometry* (Prentice Hall, 2005), 8.

only to "flat" space; non-Euclidean geometries, developed separately by Gauss, Bolyai, Lobachevsky, and Riemann, violate axiom 5. The question of whether our universe is Euclidean or non-Euclidean is a scientific one. Recent measurements indicate that space is flat (Euclidian) to a high degree.[9]

Mathematics is a great achievement of human thought. However, modern philosophies of mathematics, particularly those derived from Frege's nominalistic views, drive a wedge between mathematics and existence. This wedge is made explicit in the ZF axioms of set theory where mathematics is supposedly derived from the empty set. Yet, as Peikoff writes:

> A concept is an integration that rests on a process of abstraction. Such a mental state is not automatically related to concretes, as is evident from the many obvious cases of "floating abstractions."... If a concept is to be a device of cognition, it must be tied to reality.[10]

A *floating abstraction* is Ayn Rand's term for
concepts detached from existents.

If an individual absorbs a concept from other men without knowing the specific units the concept denotes, he or she holds a floating abstraction. Modern mathematics professors are distinct in their vehement, explicit embrace of floating abstractions. Mathematicians are not shy about expressing their Platonic views.

---

9   Ignazio Ciufolini et al., "A Test of General Relativity Using the LARES and LAGEOS Satellites and a GRACE Earth Gravity Model," *The European Physical Journal C* 76, no. 3 (March 2016): 120, https://doi.org/10.1140/epjc/s10052-016-3961-8; and Springer, "Proving Einstein Right Using the Most Sensitive Earth Rotation Sensors Ever Made," news release, May 10, 2017, https://phys.org/news/2017-05-einstein-sensitive-earth-rotation-sensors.html.

10   OPAR, 96.

Objectivism holds that concepts originate in the human mind as it interacts with existents, not in a Platonic heaven, or in physical objects, or in linguistic analysis. Mathematics, at least the mathematics that says something about the world, can be traced back to reality. Mistakes are possible, since concept formation is not automatic. The human mind must focus and perform complex processes to form a concept, differentiating each new concept from previous ones, and integrating the new concept into a meaningful thought. Correspondence with reality is sometimes hard to trace. But trace it we must: reality is the arbiter of truth. Existence does not allow contradictions; everything must conform to its own identity. If you find a contradiction in mathematics, check your premises and the logic behind your theory. Contra Bertrand Russell, an objectivist can say, with a full understanding of the metaphysical, conceptual, and epistemological source of mathematical concepts, "I know what I am talking about, and whether what I am saying is true."

THE ANSWER

Is arithmetic axiomatic? No, arithmetic is not axiomatic. The counting numbers are formed by creating words to denote an ordered list with respect to quantity. All of mathematics is derived from the counting numbers by using the Identity Axiom. Man has a choice to form a concept or not, but existence dictates whether a concept corresponds to reality or does not. Existence is axiomatic, as is consciousness and identity. Mathematics is derived from reality and, hence, is not axiomatic.

DO SPACE AND TIME EXIST?

UP IN THE AIR

A yn Rand wrote:

> The building-block of man's knowledge is the concept of an "existent"—of something that exists, be it a thing, an attribute or an action.[1]

Some readers take this passage too literally and argue that only "things," "attributes," and "actions" exist. Since time and space are none of these three items, they argue that time and space do not exist. In this view, length, width, and height are attributes and do not exist apart from the things containing the attributes.[2]

---

1 ITOE, 5.
2 The issue here is not between the theories of absolute versus relative space, whether space is invariant with respect to motion or is modified by...

To be fair, Ayn Rand confused this issue in her response to a question from Philosophy Professor A, who asked if length existed in reality:

> Length does exist in reality, only it doesn't exist by itself. It is not separable from an entity, but it certainly exists in reality. If it didn't, what would we be doing with our concepts of attributes? They would be pure fantasy then. The only thing that is epistemological and not metaphysical in the concept of "length" is the act of mental separation, of considering this attribute separately as if it were a separate thing.
>
> How would you project a physical object which had no length? You couldn't.[3]

We can extend Rand's question and answer as, "How would you project a physical object without space around it? You couldn't." David Harriman ignores this impossibility:

> Newton treated the concepts "space" and "time" as existents independent of bodies, rather than as relationships among bodies. Thus he viewed space as an infinite cosmic backdrop that exists independent of the bodies placed in it, and he claimed that this backdrop has real physical effects on the bodies that accelerate with respect to it...[Had Newton said it differently,] it would have eliminated the impossible task of trying to establish the existence of space as a supernatural pseudo-entity... The correct relational

---

...motion. The issue is whether space-time structures exist in their own right, metaphysically speaking. Nick Huggett et al., "Absolute and Relational Space and Motion: Post-Newtonian Theories," *The Stanford Encyclopedia of Philosophy*, fall 2024 ed., eds. Edward N. Zalta and Uri Nodelman, https://plato.stanford.edu/archives/fall2024/entries/spacetime-theories/.

3   ITOE, 278.

view dates back to Aristotle, who treated space as a sum of places and explained that the concept "place" refers to a relationship among bodies.[4]

The dispute with science in this statement is massive: The word "bodies" implies solid objects, ignoring fluids, gases, light, gravitational fields, electric fields, and magnetic fields. The International System of Units defines the meter as, "The distance travelled by light in vacuum in 1/299,792,458 second," whereas a second is defined as, "The duration of 9,192,631,770 periods of the radiation corresponding to the transition between the two hyperfine levels of the ground state of the caesium-133 atom."[5] Where are the "bodies" in these definitions?

At least Rand referred to "things." One can argue that light, radiation, and atoms are "things." Is the earth's gravitational field a "thing"? It certainly has a big effect on us, and it spreads throughout space![6]

To clarify the concept of space, we take Rand's advice:

When in doubt about the meaning or the definition of a concept, the best method of clarification is to look for its referents—i.e., to ask oneself: What fact or facts of reality gave rise to this concept? What distinguishes it from all other concepts?[7]

4   Harriman, *The Logical Leap*, 148–49.
5   International Bureau of Weights and Measures, *The International System of Units (SI)*, 9th ed., v3.01 (August 2024), https://www.bipm.org/documents/20126/41483022/SI-Brochure-9-EN.pdf.
6   We discuss the absurdity of Aristotle's speculative physics versus the correctness of Newton's and Einstein's experiment-based physics further in *Question 18—How does space work?* and *Question 27—Are space and time linked?*
7   The question, "What fact or facts of reality gave rise to this concept?," is known in objectivist circles as *Rand's question.* ITOE, 51.

What facts of reality give rise to the concept of space? A child first learns of space by living in it. It is helpful to repeat the following Ayn Rand quote from *Question 3—What are concepts?*

> If a child considers a match, a pencil and a stick, he observes that length is the attribute they have in common, but their specific lengths differ. The *difference* is one of *measurement*. In order to form the concept "length," the child's mind retains the attribute and omits its particular measurements...[8]

The concept length does not refer to any particular object. Length, as a concept, exists independently of the objects from which it was derived. Asked to line up in class three feet behind a fellow student, a child has no problem imagining a three-foot distance behind a classmate and can walk there without difficulty. There is no need for an object to be located three feet behind the classmate for the child to figure out where to go.

Rand's statement above that "[Length] is not separable from an entity" is especially puzzling since she is the one who explained the difference between the perceptual and the conceptual levels of consciousness. On the perceptual level, an object's length cannot be separated from the object being examined. However, on the conceptual level, "length" is not tied to any particular object. The concept "space" is formed in a child's mind, implicitly at first, by realizing the concept length can be applied to the distance between the child and any point in space, whether an object happens to be in that location or not. The axiomatic concept "space exists" is evident to young minds. As explained in *Question 9—Where does knowledge start?*, space is axiomatic and can only be defined ostensively, but the concept of space originates from a child's observation of reality:

8   ITOE, 11–12.

Space is the concept that distances exist between any two points in the universe whether objects are located at these points or not.

While the concept "length" is derived by looking at objects, once formed, length turns into the concept of distance. The length of a pencil is the same as an equivalent distance *in space*. The concepts "area" and "volume" are formed in similar ways. Geometrical concepts are not tied to particular objects but provide the framework for the space in which we reside.

Newton found he had to define an inertial coordinate system to make sense of the experiments that revealed how objects moved. An inertial coordinate system is one that is not accelerating, decelerating, or rotating. Newton's Three Laws of Motion are valid only in an inertial coordinate system. In Newton's world, it does not matter whether you fix the inertial coordinate system to the earth or to a rocket ship zooming past the earth—all that counts is that the coordinate system is not accelerating or rotating. Newton's "absolute space" is not an "entity," or even a "supernatural pseudo-entity" using David Harriman's words, because, unlike an entity, you cannot push, twist, or move space. Newton's space provides the framework by which objects can be pushed, twisted, and moved.

We note that Einstein modified Newton's view of space by showing that not all inertial coordinate systems are equivalent. To make our understanding of the universe compatible with the experimental measurements of electromagnetic fields, space and time appear to be different to two observers with two different inertial coordinate systems moving at high rates of speed with respect to one another. Further, Einstein showed that if one takes the inertial coordinate system to be at rest with a free-falling body in a gravitational field, then space, time, and matter are coupled. Thus, Einstein merged the three axiomatic concepts—entities exist, space exists, and time exists—into one. We discuss the physics of space and time more fully in *Question 27—Are space and time linked?* and *Question 28—Is gravity an illusion?* Ayn Rand argued elsewhere in the above-mentioned discussion that the attributes we perceive are not "in" the object but are the result of our minds processing the data coming into our sensory organs. In particular, she says the color of an object is not in the object:

> We perceive light vibrations as color. Therefore you would say the color is not in the object. The object absorbs certain parts of the spectrum and reflects the others, and we perceive that fact of reality by means of the structure of the eye. But then ask yourself: don't we perceive all attributes by our means of perception— including length? Everything we perceive is the result of our processing, which is not arbitrary or subjective... Everything we perceive is perceived by some means.[9]

The color red is an attribute of a red object, but color presupposes the existence of light. Without light and the sense organs in our eyes, we would not be able to see the color red. Color without light is a contradiction in terms. We note that light without space is also a contradiction in terms.

Still, I have had several conversations with space deniers. A typical conversation goes like this:

SD: Space is not a thing. Things are existents with characteristics. Only relative distances between objects exist. Space with independent attributes doesn't exist.

Me: Sure, it does. It's right in front of your face.

SD: When I look at an object, say that car on the road over there, all I see is the car, I don't see the space between my eyes and

---

9   ITOE, 279–80.

the car. There is a distance between my eyes and the entity, but there is nothing in between.

Me: But there is air between you and the car.

SD: But air just fills the space between objects. It has no dimensions of its own.

Me: Engineers model airflow around a car. One example is shown in Figure 1. They solve the fluid flow equations in the space surrounding the car. The result is the velocity, pressure, and temperature of the air in the space near the car. The lines you see in this figure correspond to the movement of air molecules around the car.

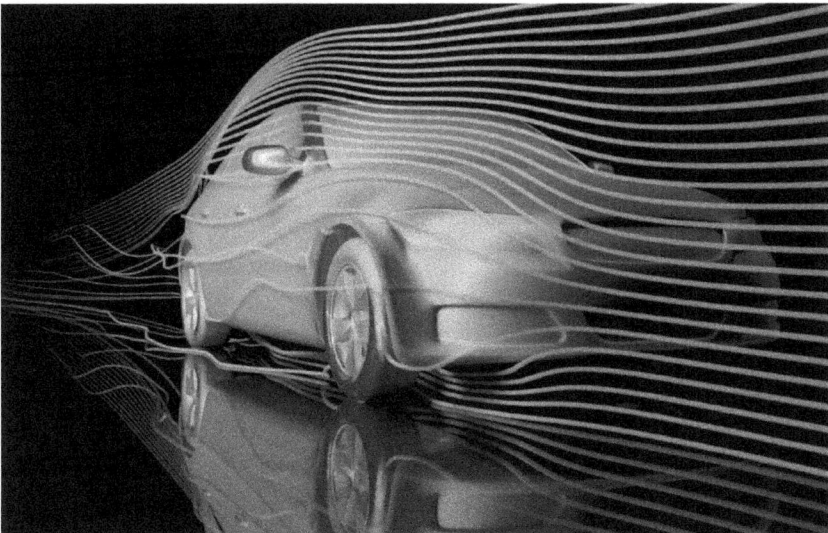

Figure 1. Computer simulation of the airflow around an automobile. The lines in this picture indicate the movement of air molecules around the car while the solid colors indicate the pressure pushing against the front of the car.[10]

10   Figure courtesy of Ansys Inc.

SD: When the wind blows, I can feel it on my face, but I don't see the movements of molecules. Admit it: You don't model the molecules of air. The lines you show in this figure are hypothetical, not real.

Me: You are correct that the lines you see in this figure are approximate. There are about 7.6×10^{23} molecules in a cubic foot of air under ordinary conditions. It is impossible to model so many molecules in any computer. To perform the simulation, space is tessellated into billions of tiny brick-shaped elements and the fluid-flow equations are modeled in each element. Combined into a large computer model, the velocity, pressure, and temperature that are computed from the model match the measured values obtained from experiments to a high degree of accuracy. But imagine if only relative distances between objects could be considered because the space between objects didn't exist. The number of relative distances between 7.6×10^{23} molecules boggles the mind!

SD: Okay. Here on Earth, air fills the spaces between objects. But there is no air in outer space.

Me: Outer space is filled with electric and magnetic fields, the gravitational field, as well as cosmic particles. You may have seen the classroom demonstration presented in Figure 2. In this demonstration, magnetic fields are made visible by placing a sheet of paper on top of a magnet and sprinkling iron filings over it.

SD: Fields are just mathematical constructs. We can only measure the forces between two objects, say between a magnet and an iron filing, but the force does not exist in the space between them. Two magnets attract or repel each other with a given force but that's the only measurable force.

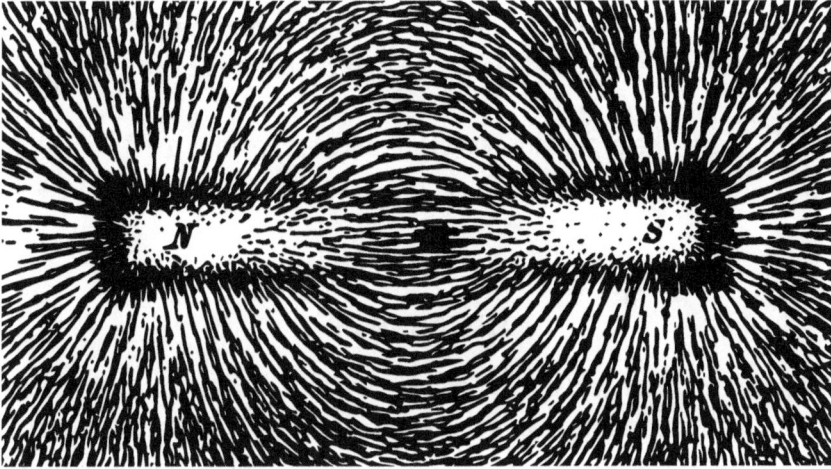

Figure 2. The magnetic field of a bar magnet revealed by dropping iron filings on paper.[11]

Me: What about light? You see the car because light leaves the car and travels *through space* to the retinas in your eyes. It takes a tenth of a millionth of a second for the light to travel the thirty feet from the car to your eyes. How does the light travel from the car to your eyes if there is no space in between?

SD: Time doesn't exist either.[12] Only events occur. One event is the light leaving the car, the other event is the light impinging on my retinas. There is no time in between. Continuous motion is an illusion. Light, or any moving object, doesn't change position until it is observed.

At this point, I give up. Space and time are axiomatic. It is not possible to convince a skeptic of a fundamental philosophic axiom other

---

11   Henry Black Newton and Harvey Nathaniel Davis, *Practical Physics* (Macmillan 1914), 242, fig. 200.

12   This is a direct quote from a conversation I had with a space and time denier.

than by pointing at it. The skeptic uses space and time throughout his or her entire life but does not recognize that he or she is doing so.

The argument "Only entities exist" entails a
primacy of consciousness point of view.

THE ANSWER

Do space and time exist? Yes, they exist. You cannot get away from them. It is impossible to conceive of an object without space around it or without the passage of time. Length is a concept derived as an attribute of objects but once formed, the concept "length" is not tied to objects. A "distance" in space is the concept "length" applied to two points separated in space.

DO GEOMETRICAL OBJECTS EXIST?

"MEASUREMENTS OMITTED" OBJECTS

Pythagoras founded a school of philosophy in Greece in 530 BCE.[1] The Pythagorean school believed that all things are made of numbers, foreshadowing such modern works as Max Tegmark's *Our Mathematical Universe*, cited earlier. Euclid codified many of the Pythagoreans' works and those of later Greek mathematicians in his *Elements*, circa 300 BCE. The ancient Greeks identified points, lines, circles, and other geometric entities as basic concepts. Yet a point has no height, length, or width; it is nothing. A line is infinitely straight and thin, a physical impossibility. And every circle you will ever draw violates perfect roundness. Since infinitely small points, perfectly straight lines, and perfect circles

---

1   Charles H. Kahn, *Pythagoras and the Pythagoreans: A Brief History* (Hackett, 2001).

are not found in existence, geometric structures conjure a mystical significance among Platonists. Our task is to demystify geometry.

The ancient Egyptians were forced to survey the lands along the Nile every year due to the annual Nile floods. Tomb paintings in the Theban Necropolis depict surveyors using knotted cords and plumb bobs to measure the lands.[2] In this technology, stakes are pounded into ground and a rope is pulled taut between them. The surveyors were called *rope stretchers* because the rope had to be stretched so it would not sag.

Surveying via rope-stretching was adopted by the ancient Greeks. The concept "line" is formed from the stretched rope by omitting measurement of the thickness of the rope; the concept "point" is formed by omitting the radius of the end-stake. Lines and points are conceptual tools that enable men to deal with "lengths" perceived in metaphysical reality, disregarding irrelevant factors such as the thickness of a rope or the radius of a stake.

As Robert Knapp explains, geometric entities are concepts of physical existents with their imperfections treated as "omitted measurements":

In Ayn Rand's terms when one measures triangles, color is an omitted measurement. A triangular object must have some color, but it may have any. The particular color doesn't matter; it does not affect one's study of shape. In the same way, when one measures triangles, any microscopic or irrelevant imperfections of the triangle are omitted measurements. If these imperfections were relevant, we couldn't count them as triangles. But if something doesn't matter, one doesn't measure it. One omits it from one's analysis.[3]

2   Walter G. Robillard et al., *Evidence and Procedures for Boundary Location*, 6th ed. (John Wiley and Sons, 2011), 282.
3   Knapp, *Mathematics Is About the World*, 36.

Geometrical concepts are the product of human consciousness. Omitting the measurements of nonsignificant aspects of shape are part of the concept-formation process. Young children differentiate shapes in many ways: by the straightness, or lack of straightness, of an object, the uniform, or nonuniform, curvature of a body, etc. The concept of a "line" is formed by differentiating the shape of thin objects from other shapes, omitting the thickness of the object, integrating the observed thin, long shapes together into a group, and assigning the word "line" to this shape. Children love to draw and are adept in drawing lines, points, and circles.

Since the thickness of a line is an omitted measurement, it can take on any value. In a physical drawing, we may choose to draw a line 0.5 millimeters thick, or 1.0 millimeter thick. On the conceptual level, however, the line thickness is an omitted measurement; a conceptual line has no thickness at all. The same process allows points to have zero length. The infinitely thin and the infinitely small are properly formed concepts even though such things do not exist in physical reality.

Since points have zero length, an infinite number of points are required to draw a line. The concept of infinity arises from geometric constructions, as it did in considering unlimited sequences.

Mathematical concepts occur in only one place: inside the human mind. Searching for geometric universals in external reality is fruitless. Hypothesizing a "Platonic heaven" to hold geometric concepts is deluded. Geometric concepts are formed in man's mind as abstractions of the shapes we see in this world.

TOPOLOGY IS BUILT WITH
"MEASUREMENTS OMITTED" GEOMETRY

Three hundred years ago, the city of Königsberg in Prussia was famous for its seven bridges connecting the mainland with two large islands in the Pregel river. Local lore maintained that a person could not walk across all seven bridges without recrossing at least one of the bridges, but no one knew why.

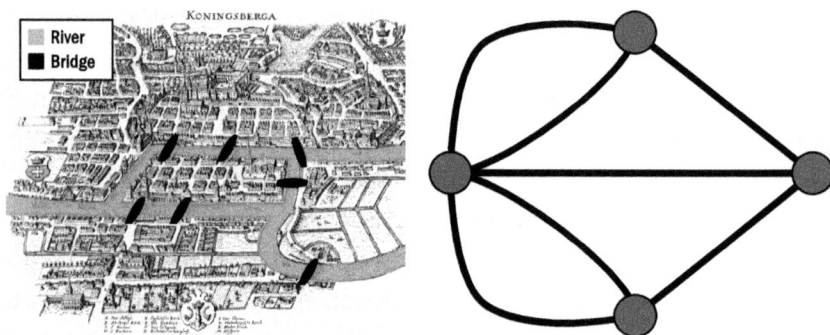

Figure 1A. The seven bridges of Königsberg. Local lore suggested that one could not cross all seven bridges without recrossing at least one of them. (1B) Euler proved the Königsberg citizens were right by converting the problem to the graph shown in this figure. Here, the upper node represents the mainland to the north of the river, the node at left the island to the left, the node at right the island to the right, and the bottom node the mainland to the south of the river.[4]

Euler solved the "seven bridges of Königsberg" problem in 1735. In the process, he developed a new branch of mathematics called *topology*. Topology deals with the relationships among geometric structures with all of their measurements omitted. The layout of the seven bridges' problem is presented in Figure 1A. From this layout, we draw the graph shown in Figure 1B. A *graph* is a set of connected lines of arbitrary lengths, called *branches*, that have the same connectivity as the layout. The points at which these branches intersect are called *nodes*. In this case, we have four nodes corresponding to the four landmasses and seven branches corresponding to the seven bridges. Euler's seminal idea was to recognize that the number of

4   Image by Mark Foskey and Booyabazooka (Wikimedia Commons, CC BY-SA 3.0).

bridges to any landmass, except for the ones chosen for the start and the finish, must be even if all bridges are to be crossed but no bridge can be crossed twice. The number of branches touching a node is called the *degree* of the node. In Figure 1B, all nodes are of odd degree: The degree of m1 is 5, and the degrees of m2, m3, and m4 all equal 3. Therefore, it is impossible to cross all seven Königsberg bridges without recrossing some of the bridges.

Euler's analysis of the Königsberg bridges' problem is the beginning of modern mathematics. Traditional mathematics focused on *quantity*; modern mathematics' interest is *structure*. Euler wrote about the bridges:

> [The result is] concerned only with the determination of position and its properties; it does not involve measurements.[5]

Quantity is a one-to-one correspondence between a counting system and concepts corresponding to existents; structure is about relationships. An objectivist looks at the Königsberg bridges and sees concepts derived from existence through measurement omission. A nominalist looks at the Königsberg bridges and sees linguistic analysis:

> If mathematics has no factual content and its theorems are meaningful, it cannot but reduce to a linguistic framework, in which some of our questions and statements concerning our experiences can be expressed, to be used in the process of verifying other statements: it only concerns the structure of our language.[6]

---

5   Leonhard Euler, "Solutio problematis ad geometriam situs pertinentis," in *Graph Theory 1736–1936*, ed. N. L. Biggs et al. (Clarenden, 1976), 3.
6   Panza and Sereni, *Plato's Problem*, 73.

This type of thinking severs mathematics from reality. To declare "mathematics has no factual content" misconstrues the nature of concepts. Concepts are formed through measurement omission. *The concepts thus formed are fact-based.* Quantity is a fact of life. Mathematics originates with the natural numbers—the numbers formed by counting the quantities of entities labeled with the same concept. Higher level mathematics is derived from the natural numbers through logic and the Identity Axiom.

> Language is far more than structure: True statements, mathematical or otherwise, speak about relationships in reality; language unrelated to reality is gobbledygook.

Euler was correct in saying, "[The Königsberg bridge problem] does not involve measurements." However, he did not appreciate the central role of measurement omission in concept formation. Measurement omission is the rule in mathematics, not the exception.

An Aristotelian realist looks at the Königsberg bridges and searches for universals:

> Once the existence of the science-of-quantity and science-of-structure theories is noticed, some obvious questions arise. Are they really the same theory, with structure being just a modern understanding of what was previously called quantity?... The position that will be argued for here is that quantity and structure are different sorts of universals, both real.[7]

Quantity and structure do differ—quantity is measurements kept; structure is measurements omitted. However, the idea that universals are real and exist independently from human consciousness

7   Franklin, *An Aristotelian Realist Philosophy of Mathematics*, 33–34.

is bizarre. No one has ever seen a "universal" embedded in entities apart from man's mind. The measurement-retaining and the measurement-omitting processes required to form a concept occur when man is thinking. They are not to be found in existence outside the human mind.

THE ANSWER

Do geometrical objects exist? No, the shapes of objects are physical, but geometrical concepts such as the point, the line, and the circle exist only in man's mind. Geometric entities are concepts derived from shapes in existence with their measurements omitted.

QUESTION 17

DOES A LINE REALLY CONTAIN AN INFINITE NUMBER OF POINTS?

GEOMETRY AS NUMBERS

A remarkable event occurred in 1637. René Descartes put axes on space, put numbers on the axes, and united arithmetic and geometry. Prior to 1637, arithmetic and geometry were regarded as different spheres of knowledge. Arithmetic was used to count discrete things; geometry was used to measure continuous distances. Descartes's invention of the Cartesian coordinate system made mathematics whole.[1]

---

1   "The person who is popularly credited with being the discoverer of analytic geometry was the philosopher René Descartes (1596–1650), one of the most influential thinkers of the modern era." Roger Cooke, *The History of…*

The Cartesian coordinate system assumes perfectly perpendicular axes. Perfect points, lines, and circles do not exist in metaphysical reality; these geometric concepts are conceptual abstractions. However, the concept of perpendicularity is formed by studying space, and space is given metaphysically. The three independent space variables are *metaphysical*.[2]

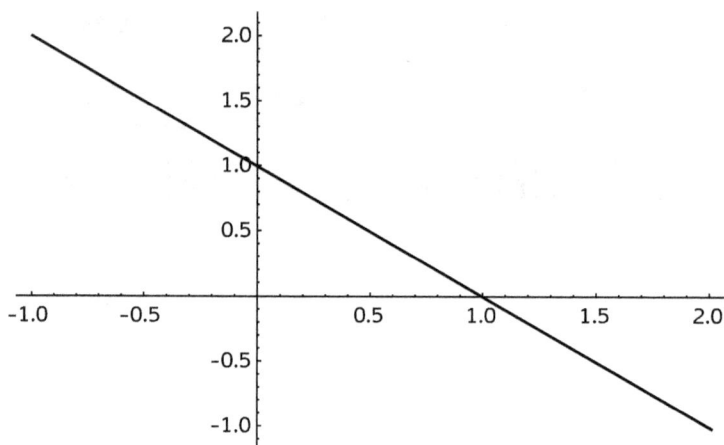

Figure 1. Plot of the equation $x + y = 1$ on the two-dimensional Cartesian coordinate system.

In two dimensions, the Cartesian coordinate system is drawn with a horizontal axis and a vertical axis, as shown in Figure 1.[3] Evaluating the quantity:

...*Mathematics: A Brief Course* (Wiley-Interscience, 1997), 326. Pierre de Fermat's notes on analytical geometry circulated in Paris prior to Descartes's work, but Descartes beat Fermat to publication. See Victor J. Katz, *A History of Mathematics: An Introduction*, 2nd ed. (Addison Wesley Longman, 1998).

2   Einstein's discovery of space-time does not alter the metaphysical basis of space and time. Space-time extends our knowledge of the properties of space and time but does not alter their existence.

3   Three orthogonal coordinate axes represent three-dimensional space.

$$x + y = 1$$

at $x = 0$ gives $y = 1$ and at $y = 0$ gives $x = 1$. Since the relationship between x and y is linear, beginning at $x = 0$ and adding any quantity to x subtracts the same quantity from $y = 1$:

$$y = 1 - x$$

For example, at $x = \frac{1}{4}$, we obtain $y = \frac{3}{4}$. The result is the straight line plotted in Figure 1. The line segment between $x = 0$ and $x = 1$, and the lines $x = 0$ and $y = 0$ provided by the axes, form the triangle shown in Figure 1.

In an Aristotelian realist's mind, arithmetic and geometry are incompatible: Arithmetic counts discrete entities, geometry measures continuous distances. As Franklin notes:

> Therefore geometry, and continuous quantity in general, are in some fundamental sense richer than arithmetic and not reducible to it via choice of units.[4]

The problem for the realist is that numbers are "in" entities, but infinity is a concept and therefore cannot exist. Mathematical operations, such as functional analysis and calculus, that are defined over an infinite set of points along a line are impossible for a realist.

TO INFINITY AND BEYOND

Consider the line segment from $x = 0$ to $x = 1$. We can place point after point on this line segment and never fill it up. A point has zero length, so placing point after point takes up no length. There is space between every point we place, even as we place an unlimited number

---

4   Franklin, *An Aristotelian Realist Philosophy of Mathematics*, 44.

of points along the line segment. We need a new concept to fill in the line. We must introduce an infinite number of points between every point we place, points so numerous that we will never be able to write them to full precision using decimal arithmetic.

To understand this concept, note that, when written in decimal form, every fraction generates a repeating set of digits. For example, 1/3 written in decimal form is:

$$1/3 = 0.3333333333333333\ldots$$

where the...indicates that the 3s go on forever. As another example, consider the fraction 10/7, the value proposed by the Indiana legislature in the Indiana Pi Bill for the quantity √2:

$$\text{Pi Bill value of } \sqrt{2} = 1.42857142857142857\ldots$$

Here, the sequence of digits 142857 repeats forever. We call a number that provides a repeating pattern of digits a *rational number*. A number that lacks a repeating pattern is called an *irrational number*.

Consider a number generator that provides one of the ten single-digit integers {0, 1, 2,..., 9} at random. Running this random-number generator twice provides two integers, d_1 and d_2. Writing each integer as a digit in a decimal expansion provides 100 values x_i from $x = 0.00$ to $x = 0.99$:

$$x_i = 0.\, d_1\, d_2$$

d_1 has 10 values; d_1 must equal d_2 for x_i to contain repeated digits. Thus, the two-digit decimal contains 10 numbers with repeating digits and 90 numbers with non-repeating digits. Let R be the ratio of repeating digit numbers to non-repeating digit numbers. In the two-digit case, we find that:

$$R_2 = 10/90 = 0.11111111\ldots$$

Now run the random-number generator four times to provide four integers d_1, d_2, d_3, and d_4:

$$x_i = 0.d_1\, d_2\, d_3\, d_4$$

In this case, d_1 has 10 values, as before, and these values may repeat with d_2, d_3 and d_4. In addition, the two-digit combination d_1 and d_2 may repeat, $d_3 = d_1$ and $d_4 = d_2$, and there are 100 such numbers. Thus, in the four-digit case, there are 110 numbers with repeating digits and 10,000−110 = 9,890 numbers with non-repeating digits. In this case, the ratio of repeating digit numbers to non-repeating digit numbers is:

$$R_4 = 110/9{,}890 = 0.01112235\ldots$$

The ratio R becomes smaller as we double the number of digits from two to four.

The ratio R becomes smaller each time we repeat this process, doubling the number of digits from four to eight, then from eight to sixteen, and so on. Although we could never carry out this process forever, we can ask the conceptual question: What happens if we let the number of doublings go to infinity? In that case, R goes to zero.

> The chance of writing a hundred digits at random and coming up with a repeating pattern is unlikely compared to creating a set of uncorrelated digits.

Since repeating pattern numbers are fractions written in decimal form—recall we called these *rational numbers*, and there is no limit to the number of fractions we can form, the number of rational numbers is infinite. The argument above reveals that the number of non-repeating pattern numbers known as *irrational numbers* exceeds any finite count. The set of irrational numbers vastly outweighs the set of rational numbers.

Georg Cantor discovered the relationship between rational numbers and irrational numbers in the nineteenth century.[5] Prior to Cantor, infinity was a vague concept, such as the understanding that the decimal expansion of 1/3 repeats forever. After Cantor, infinity had structure and richness. Cantor turned the Descartes geometric axis into the number line.[6]

Cantor's proof was simple. Create a numbered list of random numbers in the interval from 0.0 to 1.0:

$$s_1 = 0.d_{1,1}\ d_{1,2}\ d_{1,3}\ d_{1,4}\ d_{1,5}\ d_{1,6}\ \cdots$$
$$s_2 = 0.d_{2,1}\ d_{2,2}\ d_{2,3}\ d_{2,4}\ d_{2,5}\ d_{2,6}\ \cdots$$
$$s_3 = 0.d_{3,1}\ d_{3,2}\ d_{3,3}\ d_{3,4}\ d_{3,5}\ d_{3,6}\ \cdots$$
$$s_4 = 0.d_{4,1}\ d_{4,2}\ d_{4,3}\ d_{4,4}\ d_{4,5}\ d_{4,6}\ \cdots$$
$$s_5 = 0.d_{5,1}\ d_{5,2}\ d_{5,3}\ d_{5,4}\ d_{5,5}\ d_{5,6}\ \cdots$$
$$s_6 = 0.d_{6,1}\ d_{6,2}\ d_{6,3}\ d_{6,4}\ d_{6,5}\ d_{6,6}\ \cdots$$
$$\vdots \qquad\qquad \vdots$$

where s_i is the i'th random number and $d_{i,j}$ is the j'th digit in the i'th random number. The number of random numbers thus produced equals the number of counting numbers 1, 2, 3,.... Next increment the value of each diagonal digit $d_{i,i}$ to the next value:

$$\text{If } d_{i,i} = 0, \text{then } d'_{i,i} \text{ becomes } 1$$
$$\text{If } d_{i,i} = 1, \text{then } d'_{i,i} \text{ becomes } 2$$
$$\vdots$$
$$\text{If } d_{i,i} = 9, \text{then } d'_{i,i} \text{ becomes } 0$$

where $d'_{i,i}$ is the new value of the i'th row diagonal entry. Now form a new number using the new values of the diagonals:

5   Phillip E. Johnson, "The Genesis and Development of Set Theory," *The Two-Year College Mathematics Journal* 3, no. 1 (Spring 1972): 55–62.
6   Joseph Warren Dauben, *Georg Cantor: His Mathematics and Philosophy of the Infinite* (Princeton University Press, 1979).

$$s_{new} = 0.\, d'_{1,1}\ d'_{2,2}\ d'_{3,3}\ d'_{4,4}\ d'_{5,5}\ d'_{6,6} \cdots$$

s_{new} cannot be equal to s_1 because it differs from s_1 in the first decimal place, it cannot be equal to s_2 because is differs from s_2 in the second decimal place, and so on throughout the list. Thus, a number s_{new} exists that differs from all the numbers on the list. Since each row on the list corresponds to a counting number, 1, 2, 3,..., the number of real numbers must be greater than the number of counting numbers. It is easy to show that the number of counting numbers is the same as the number of rational numbers, i.e., the number of positive or negative fractions. Therefore, the set of real numbers, both rational and irrational, is not countable.[7]

The inability to pair integers with real numbers reveals two types of infinity: a "countable infinity," represented by natural numbers, and an "uncountable infinity," represented by real numbers.

Ludwig Wittgenstein could not tolerate Cantor's proof. Cantor's mathematics is "utter nonsense," "laughable," and "wrong," Wittgenstein wrote.[8] He failed to realize that the natural numbers are derived from reality and that all other mathematical concepts are derived from these numbers by using the Identity Axiom.

High schools teach Cantor's proof today because it is so easy. How did a clever man such as Ludwig Wittgenstein fail to understand this

7   The observant reader will have noticed that the above logic fails if two or more random numbers are the same. It turns out that the probability of two numbers in the list being the same decreases exponentially as n gets larger. Thus, the probability of two random numbers ri and rj being the same reduces to 0 as n increases toward infinity. See Geoffrey Hunter, *Metalogic: An Introduction to the Metatheory of Standard First Order Logic* (University of California Press, 1973).

8   Victor Rodych, "Wittgenstein's Philosophy of Mathematics," *The Stanford Encyclopedia of Philosophy*, spring 2018 ed., ed. Edward N. Zalta, https://plato.stanford.edu/archives/spr2018/entries/wittgenstein-mathematics/.

simple proof? The answer is Wittgenstein's primacy of consciousness mode of thinking. Mathematics must correspond to my preconceived notions, Wittgenstein apparently thought, not to what reason and reality dictate. However, reality is what it is. Mathematical proof must be derived from reality, not idle speculation.

THE ANSWER

Does a line really contain an infinite number of points? Yes, points along a line are so many you cannot count them. Further, the concept of infinity has a structure. Some infinities contain more elements than other infinities. The proof is simple but disbelieved by those who put their preconceived notions ahead of logical arguments.

HOW DOES SPACE WORK?

SUPERSTITION VERSUS "NATURAL LAW"

Modernism became ascendant with the publication in 1687 of Isaac Newton's *The Principia: Mathematical Principles of Natural Philosophy*. Prior to *The Principia*, superstitions, daemons, and gods ruled men's thoughts. People believed that natural events like lightning strikes were the work of gods. They also believed that nonliving things, such as rocks and rivers, had goals and desires. Moreover, living men imagined what happened after death. The publication of *The Principia* changed the relationship between man and his thoughts of existence. In the new paradigm, "natural laws" governed material objects. Philosophy, the study of the fundamental nature of knowledge, reality, and existence, divided into two branches: those who stuck with a mystical view of life and of the universe, and those who sought natural causes to observed experiences. For naturalists, physics became the foundation of a new philosophy. As David Wallace notes, "...physics tells us how to do a metaphysics that is scientifically informed."[1]

Except that life is not that simple. Concept formation requires a volitional act inside a human mind. While concretes exist separate from man's mind, the concepts man forms from these concretes do not. Concepts are man's way of understanding reality, but they are a human product with human characteristics. Doing "a metaphysics that is scientifically informed" is an epistemological process, as well as a metaphysical one.

Man's method of cognition involves the specific process discussed in *Question 2—What is consciousness for?* And *Question 3—What are concepts?* Key steps in this process are to focus on the essential attributes of existents, differentiate these attributes from all others, omit the measurement of these attributes, integrate the new concept into the individual's existing knowledge base, and label the new concept with a word. This process is a prerequisite for all scientific study. Science, as an element of human knowledge, can be obtained only through human means. Immaculate knowledge, knowledge apart from living beings, does not exist.

Men learn from each other, but everyone must differentiate every new idea, whether self-generated or acquired from others, from the ideas he or she has already learned. A child's range of knowledge is limited and must focus on basic concepts. An adult may have a vast range of knowledge and need to integrate a variety of complex concepts. However, in every case, an individual human mind must differentiate each idea from others and integrate it. A "knowledge pill," sometimes used as a literary device in science fiction where an individual receives knowledge through osmosis, does not exist.

---

1   David Wallace, *Philosophy of Physics: A Very Short Introduction* (Oxford University Press, 2021), 3.

The "collective brain" is another nonsense concept. Concepts originate with the differentiation/measurement-omission/integration/naming process described above. These steps are necessary for the formation of any concept. The only place this process occurs is in the individual human brain. It does not happen in a disjointed "collective brain." While we can speak of the common knowledge shared by a society, every individual who holds it processes individually every idea. The common knowledge of a society is just the aggregate of the knowledge held by the individuals in that society.

Leonard Peikoff writes, "...thinking, to be valid, must adhere to reality."[2] Reality imposes the following requirements:

1. Valid thoughts must originate in existence. Man must look outward, at existence, to understand that which exists. Flights of fancy, unconnected to reality, are not knowledge.
2. Valid thoughts must correspond to that which exists. Science is not a mental game but the pursuit of ideas that describe reality.
3. Valid thoughts must be self-consistent. Everything we know is interrelated. If two ideas contradict each other, you know at least one of them is wrong.

The first two of these requirements, identified by Ayn Rand, prompted Rand to call her philosophy objectivism. Man derives objective knowledge from things that exist through his means of concept formation. Ideas that do not correspond to things that exist are not real. An idea unprocessed by a human mind is a contradiction in terms. The first of these requirements rules out subjectivism, the philosophy that ideas come from God or originate in man's "collective mind;" the second rules out intrinsicism, the philosophy that ideas exist in things independently from man.

---

2   OPAR, 110.

The third requirement was recognized by Aristotle who formulated the rules of logic. Aristotle's three rules of logic are:

1. The Law of Identity: *"Whatever is, is."*
2. The law of noncontradiction: *"Nothing can both be and not be."*
3. The law of excluded middle: *"Everything must either be or not be."*[3]

Armed with these three laws, and cognizant of man's method of concept formation, we are ready to determine whether the universe is predetermined or unpredictable.

SPACE, TIME, AND MATTER

Metaphysics begins with the axiom of existence and its corollaries. The three physics-based corollaries are: space exists, time exists, entities exist. It follows that physics is a scientific study of the properties of space, time, and matter. Yet, in answering *Question 15—Do space and time exist?*, for many people, space and time appear to be unreal. Reality, relationists argue, are things you can see and touch.[4] The rock you hold in your hand is real, they say, as well as its power to smash a nut, but the space between you and the nut does not have an independent identity.

The relationist view of space and time originated with classical Greek philosophy. Circa 500 BCE, the Greek philosopher Parmenides argued that motion is impossible because motion requires moving

---

3   Rand paid tribute to Aristotle by naming the three parts of *Atlas Shrugged* after Aristotle's laws. Aristotle, *Metaphysics*, Book IV, parts 4 and 7, trans. W. D. Ross (1908).

4   Relationism is the view that all there is in the world is matter; the idea that space as a thing separate from matter is called *substantivalism*.

an object into the void, the "void" is nothing, and nothing does not exist.[5] Without space and time, nothing can move or change.

Aristotle used teleological concepts to explain physical phenomena.[6] Release a rock in midair, Aristotle said, and the rock falls of its own volition.[7] Aristotle conjectured that the goal of the rock is to seek the center of the universe. The Greek philosopher Empedocles hypothesized that material objects are composed of mixtures of four fundamental elements: earth, air, fire, and water.[8] Aristotle assigned a hierarchy of "natural" motions to these elements: Rocks fall in water, water falls in air, air bubbles up in water, and a flame rises in the air. According to Aristotle, the goal of both water and of earth is to seek the center of the universe, but earth is more fervent in this downward-seeking goal than is water. Similarly, the goal of both air and fire is to move upward, away from the center of the universe, with fire more eager in this goal than air.[9]

Aristotle's "displacement" view of falling and rising objects caused him to write that the speed of a falling body is proportional to its weight and inversely proportional to the density of the fluid in which it is falling.[10] A vacuum is impossible, Aristotle said, because the speed of a falling body in a vacuum would be infinite.

---

5   Betrand Russell, *A History of Western Philosophy* (Simon & Schuster, 1945), 69.

6   Harry Binswanger, *The Biological Basis of Teleological Concepts* (Tof, 1990).

7   While Aristotle was the most reality-focused philosopher of the ancient world, he did not perform experiments. For example, he wrote that the speed of a falling body was proportional to its weight. It would take nearly two thousand years for Galileo to perform the experiment and prove Aristotle wrong. See Carlo Rovelli, "Aristotle's Physics: A Physicist's Look," *Journal of the American Philosophical Association* 1, no. 1 (2015): 23–40, https://doi.org/10.1017/apa.2014.11.

8   Peter Kingsley, *Ancient Philosophy, Mystery, and Magic: Empedocles and Pythagorean Tradition* (Oxford University Press, 1995).

9   Tim Maudlin, *Philosophy of Physics: Space and Time* (Princeton University Press, 2012), 1.

10   Rovelli, "Aristotle's Physics."

To account for the motion of the stars, Aristotle proposed a fifth element, the aether. The goal of aether is circular motion around the center of the universe, the path that he believed stars took in the heavens. The five "wandering stars" of the ancient world (Mercury, Venus, Mars, Jupiter, and Saturn) deviate from this path. For two millennia, astronomers drew small circles with centers located on the perimeters of larger circles, attempting to use circles drawn upon the perimeters of circles, to replicate the motions of planets. Even Copernicus required the planets to follow circular paths, although he did place the sun at the center of the universe. It was only with Kepler's publication of the first law of planetary motion in 1609 that Aristotle's heavenly circular motion was replaced with the ellipse as the correct planetary path.

It is rare to find a clearer example of errors in concept formation. Yes, the sun, moon, and the stars take circular paths across the sky, but it is the earth that rotates, not the universe. Yes, a feather floats gently down through the air with a limited velocity, but heavy objects all fall at the same accelerating rate. Aristotle's physics appears non-sensical to us today, but his ideas prevailed for nearly two thousand years. How could one of the smartest people who ever lived, a person who discovered the rules of logic, be so profoundly wrong? Where was Aristotle's mistake?

The key to science, indeed the key to all clear thinking, is to place existence and consciousness in their proper places. Consciousness is an observer of existence, not its creator. Ideas such as "the goal of earth is to seek the center of the universe" or "God placed the Earth at the center of the universe" introduce arbitrary teleological and/or theological elements into "the laws of physics."[11]

---

11   The word "law", used to label a fundamental physical principle, is a remnant of theological thinking. A "law" is a commandment issued by a consciousness. Calling the foundational principles of physics "laws" implies that these...

Incidental observation is one thing, performing an experiment is another. An experiment requires thought and an expectation of some physical effect. An experiment without a concept of what is being measured is of no use at all. The scientific method requires forming a hypothesis, creating an experiment to test that hypothesis, and determining whether the measured results confirm or invalidate the hypothesis.

Isaac Newton's bucket experiment illustrates this process well. Maudlin calls Newton's bucket experiment "one of the most powerful and compelling experiments in the history of physics."[12] Newton hypothesized that space is an absolute thing, and motion is movement regarding this absolute space, but how did he show this?

Aristotle had argued for a "circular inertia," the idea that, without interference, objects would circle the center of the universe forever. Galileo, despite his renaissance role as a pioneering experimental physicist, supported Aristotle's view. Galileo set up two ramps, one running down an incline, the other running up. He let a ball run down the first ramp and observed that the ball ran up the second ramp to almost the same level it had when he had let go of it. Galileo concluded, correctly, that, in the absence of friction, the ball would reach the same height as it had at its start. He hypothesized, correctly, that a ball sent rolling on a frictionless, spherical earth would roll around the earth forever. From this, he concluded, this time incorrectly, that the natural motion of a ball was to circle the earth at a constant height.

…"laws" were established by a Creator. Yet, existence exists independently of any consciousness, consciousness is awareness of existence not its creator, so the word "law" reverses the roles of consciousness and existence. The proper terminology for "laws of nature" is "properties of nature." Nevertheless, the use of the word "law" is so well established in physics that we shall continue to use it with the understanding that each "law" we cite is an observed property of nature, not a divine commandment.

12   Maudlin, *Philosophy of Physics*, 23.

Newton set out to prove Aristotle and Galileo wrong. Take a bucket of water, he said, and hang it from a rope that has been well twisted. Before you release the bucket, neither the bucket nor the water inside it is rotating, and the surface of the water is flat. The weight of the water and the bucket will pull down on the rope and cause it to unwind. As the rope unwinds, the bucket rotates; however, the water in the bucket remains still. With the progression of time, the viscosity of the water will cause the water inside the bucket to rotate. Newton writes:

> [After some time] the water will...begin sensibly to revolve, and recede by little and little from the middle, and ascend to the sides of the vessel, forming itself into a concave figure (as I have experienced), and the swifter the motion becomes, the higher will the water rise, till at last, performing its revolutions in the same times with vessel, it becomes relatively at rest in it. This ascent of the water shows its endeavor to recede from the axis of its motion; and the true and absolute circular motion of the water, which is here directly contrary to the relative, becomes known and may be measured by this endeavor...[13]

This experiment is so simple and the physical effects it shows are so familiar that anyone could have done it. Yet, no one before Newton knew what to look for. Newton set up the bucket experiment to study the nature of space: Are the forces caused by rotation because of the relative rotation of the surrounding bodies, or are they because of rotation with respect to space itself? Newton's conclusion was that space exists independent of the material objects that inhabit it.

---

13   Isaac Newton, *Mathematical Principles of Natural Philosophy*, trans. based on 3rd Latin ed., vol. 1 (1729), 11.

To further clarify this point, Newton performed a thought experiment, anticipating and ultimately creating the space age. Let two globes be in outer space, far away from other material objects. Connect these globes by a cord and set the globes spinning about their common center of gravity. There will be a force on the cord dependent on the speed of rotation and the masses of the globes. We measure the speed of rotation relative to absolute space. If you cut the cord, the globes will no longer follow a circular path: Each globe will continue moving at the speed and in the direction it was moving when the cord was cut.

From these experiments, Newton established three things:

1. A rotating bucket does not produce an outward (centrifugal) force in the water when the water is not rotating.
2. The force pushing the water outward depends on *the speed of rotation of the water with respect to space.*
3. A cord holding two rotating globes in space together is under tension, placing the same inward (centripetal) force on each of the globes.

Space is axiomatic, but its properties are not. Through these two experiments, Newton concluded that the circular motion of an object through space requires that a force apply to the object. Water spinning in a bucket pushes away from the center, whether the bucket is spinning or not; cut the cord between the system of two rotating globes and the globes no longer circle one another.

The surface of the water is level only when the water is not rotating with respect to space; zero tension on the cord occurs only when the globes are not spinning with respect to space.

Newton's First Law of Motion describes the nature of motion in absolute space:

Law 1. Every body perseveres in its state either of rest or of uniform motion in a straight line, except insofar as it is compelled to change its state by impressed forces.[14]

To explain the outward force on the water and the inward force on the globes, Newton invented a new concept: *mass*. Mass is the constant of proportionality between how much force is applied to an object and the acceleration it experiences. Mass is there, but hidden, in Newton's Second Law of Motion:

Law 2. A body acted upon by a force moves in such a manner that the time rate of change of momentum equals the force.[15]

Momentum is mass times velocity. In most situations, mass remains constant as a force is applied. In these situations, "the time rate of change of momentum" is the same as "the product of the time rate of change of velocity and the mass." Though Newton did not know that mass increases as the velocity of an object approaches the speed of light, Newton's formulation allows for the possibility of a changing mass. The Second Law of Motion is valid even when relativistic effects are included.

In *Philosophy of Physics*, Maudlin writes:

...physics is evidently in the business of postulating unobservable entities in service of explaining observable behavior. The postulation is always risky, but, as the atomic hypothesis illustrates,

---

14   Maudlin, *Philosophy of Physics*, 4–5.
15   Newton, *Mathematical Principles of Natural Philosophy*.

the risk can sometimes pay off handsomely. Newton knew that absolute space and time are not, in themselves, observable, but he also explained how postulating them could help explain the observable facts. Why is this any worse than postulating atoms?[16]

Mass is another one of those unobservable entities. Time, length, and force can be felt. Mass cannot. Mass is the constant of proportionality between force and the acceleration a body experiences. The force you apply to an object and the consequent acceleration may vary, but the mass stays constant (unless you accelerate to relativistic velocities). As a result, mass, the constant of proportionality, takes center stage in the study of physics. Mass is a higher-level concept, never to be measured directly, but central to our understanding of the galaxy.

The First Law of Motion provides the relationship between the motion of an object and its location. If an object is located at $x_0 = 0$ at time $t_0 = 0$, then in the absence of impressed forces, its location at time t will be:

$$x = vt$$

where v is the velocity of the object. The Second Law of Motion extends the relationship between motion and location to accelerating objects.

Objectivist literature presents a different view of motion:

The concept of "location" arises in the context of entities which are at rest relative to each other. A thing's location is the place where it is situated. But a moving object is not at any one place— it is in motion. One can locate a moving object only in the sense of specifying the location of the larger fixed region through

---

16   Maudlin, *Philosophy of Physics*, 46.

which it is moving during a given period of time. For instance: "Between 4:00 and 4:05 p.m., the car was moving through New York City." One can narrow down the time period and, correspondingly, the region; but one cannot narrow down the time to nothing in the contradictory attempt to locate the moving car at a single, fixed position.

If it is moving, it is not at a fixed position. The law of identity does not attempt to freeze reality. Change exists; it is a fact of reality. When a thing is changing, that is what it is doing, that is its identity for that period. What is still is still. What is in process is in process. A is A.[17]

This argument is on the primacy of the consciousness side of the existence/consciousness divide. In a primacy of existence universe, a ball rolling down an incline is always located somewhere—it does not leave the universe just because it is moving—and Newton's Laws of Motion always provide both the ball's velocity and location. It may not be possible to compute all the forces acting on the ball, and hence always compute its exact location as it rolls down the ramp, but that's an imprecision in our thinking, not in reality. In a primacy of a consciousness universe, our failure to pinpoint the exact location of a moving ball suggests the ball's metaphysical state shifts when it is still and when it is moving. A ball rolling down a ramp may cross the halfway point on the ramp, according to this view, but it was never at this point because it did not stop there. Of course, the ball does not care—it touches the halfway point—and all the other points along its path—and rolls on.

In the spinning globes thought experiment, the rope pulls on globe A with the same force as it pulls on globe B. This is because

17   Harry Binswanger, "Q & A Department: Identity and Motion," *The Objectivist Forum*, December 14, 1981, https://courses.aynrand.org/lexicon/motion/.

the tension in the single rope transmits the force to each globe. This observation led Newton to formulate the Third Law of Motion, which states that forces always cancel out:

> *Law 3.* If two bodies exert forces on each other, these forces are equal in magnitude and opposite in direction.[18]

Archimedes originated the principle "forces balance in static systems" circa 240 BCE. Archimedes studied and explained the laws of levers and pulleys. He noticed that if one arm of a lever was half as long as the other, a weight twice as heavy must be placed on that side to balance a weight half as heavy on the other side. In a famous anecdote, Archimedes ran naked down the street shouting, *"Eureka!"* ("I've found it!") when he discovered the principle of buoyancy. Archimedes had been taking a bath and realized that the weight of his body balanced the weight of water pushed up higher in the tub as he floated in the tub. Newton extended this "forces balance" principle to dynamic systems with his third law.

The earth and the moon revolve around their common center of gravity, but no cord connects these two globes. Newton introduced a force he called gravity (from the Latin *gravitate*, meaning heavy) that would act without a cord to hold the moon and the earth together. Newton proposed that gravity decreases from the earth in the same way that light spreads out from a light source. The light shining on a page twenty feet from a light bulb is one quarter as bright as that shining on a page ten feet away. This is because the light from an omnidirectional source spreads out equally in all directions. The surface area of a sphere centered on the source, the area over which the light is distributed as it radiates away from the source, grows as the radius of

---

18   Newton, Isaac. *Mathematical Principles of Natural Philosophy.* (1729 English translation based on 3rd Latin edition.) p. 19.

the sphere squared. Gravity, Newton said, decreases with the inverse square of the distance from the source, just as the intensity of light decreases with the inverse square of the distance from the source. Measurements of gravity, light, and other conserved quantities show that space is three-dimensional. The total gravitational field emanating from a body is fixed by its mass, but this field is spread over a larger spherical surface area as you move away from the body.

The inverse square relationship is a property of three-dimensional space.

Newton's Law of Universal Gravitation states:

Gravitation. A gravitational force exists between two bodies proportional to the product of their masses divided by the square of the distance between them.[19]

Proof of gravity's inverse square relationship with the distance came by calculating the orbits of planets subjected to such a gravitational field. Newton showed that the resulting orbits were elliptical, as had been established by Kepler using measurements taken by Tycho Brahe. The sequence of events here is instructive. Tycho Brahe made accurate measurements of the positions of the planets from 1588 until his death in 1601. Using Tycho's astronomical data, Johannes Kepler deduced his Three Laws of Planetary Motion, published between 1609 and 1619. The basic concept underlying these laws is that the planets move along elliptical orbits with the sun at one of the two foci. In 1687, Newton showed that Kepler's laws are a consequence of the Three Laws of Motion and the Law of Universal Gravitation.[20]

---

19   Newton, *The Mathematical Principles of Natural Philosophy, Volume II*, 220.
20   Carl Sagan and Ann Druyan, *Comet* (Random House, 1997), 52–58.

For Newton, the natural motion of an object is a straight line, and all points in space are equivalent. Natural motion, according to Aristotle and Galileo, was circular about the center of the earth, or in straight lines toward or away from the central point. Newton explained that Galileo's frictionless sphere rolls along the surface of the earth because the force of gravity keeps it from traveling in a straight line.

Mass appears twice in Newton's equations: once as the constant of proportionality between force and acceleration; a second time as the origin of the gravitational force—as we explain in answering *Question 27—Are space and time linked?* In the twentieth century, Einstein used the equivalence of these two definitions of mass to derive the theory of general relativity.

THE ANSWER

How does space work? Space works through inertia—objects move through space in straight lines at constant velocities unless acted on by a force. Newton invented a new, not directly observable quantity called mass to explain inertia and the observed motion of bodies. In Newton's use, all points in space and time are equivalent—no point in space is special or provides the center of the universe—and motions are undistorted by the relative velocities of two objects or by the gravitational field they experience. Einstein showed that this latter statement is not true when the relative velocities of the objects and/or the gravitational fields the objects experience are high. We explain this further in *Question 27—Are space and time linked?* and *Question 28—Is gravity an illusion?*

DO WE LIVE IN A CLOCKWORK UNIVERSE?

NEWTONIAN MECHANICS AND DETERMINISM

Philosophy swung from "objects move by the will of the gods" in the seventeenth century to "all movement is predetermined" in the nineteenth century, leading to the theory of a *clockwork universe*. Newtonian mechanics showed that given the position, momentum, and forces acting on a single object, its position and velocity could be determined forevermore into the future. It was a small leap, though an incorrect one, to surmise that the same result would hold for multiple objects. The French mathematician Pierre-Simon marquis de Laplace wrote in 1814:

> We may regard the present state of the universe as the effect of its past and the cause of its future. An intellect which at a certain moment would know all forces that set nature in motion, and all positions of all items of which nature is composed, if this intellect

were also vast enough to submit these data to analysis, it would embrace in a single formula the movements of the greatest bodies of the universe and those of the tiniest atom; for such an intellect nothing would be uncertain and the future just like the past would be present before its eyes.[1]

Prior to Laplace, the argument for determinism had been a theological one: Since God is omniscience and omnipotent, He knows and controls all things past, present, and future. All events are preordained, according to this view, by a monotheistic deity.[2] Laplace secularized this concept, arguing that Newton's laws could determine the future provided sufficient information on initial positions and evolving forces were available. Causal determinism, the view that "every event is necessitated by antecedent events and conditions together with the laws of nature,"[3] gained favor among physicists and philosophers alike. Critics argued that free will, the view that people can choose between different courses of action, was incompatible with this clockwork universe.[4]

Determinism reverses the roles of consciousness and existence.

The determinist says, "I may not predict what will happen in the future, but reality *knows* what will happen." The fact is things

1   Pierre Simon Laplace, *A Philosophical Essay on Probabilities*, trans. Frederick Wilson Truscott and Frederick Lincoln Emory (John Wiley & Sons, 1902), 4.

2   Anne Jordan et al., *Philosophy of Religion for A Level*, OCR ed. (Oxford University Press, 2004).

3   Carl Hoefer, "Causal Determinism," *The Stanford Encyclopedia of Philosophy*, winter 2009 ed., eds. Edward N. Zalta and Uri Nodelman, https://plato.stanford.edu/archives/sum2024/entries/determinism-causal/.

4   James A. Harris, *Of Liberty and Necessity: The Free Will Debate in Eighteenth-Century British Philosophy* (Clarendon Press, 2005).

do what they do, but things know nothing. The world obeys natural law; only natural law does not say that existents are determined for all time just because of the way things are at some initial time *t*. According to Hoefer:

> Determinism is true of the *world* if and only if, given a specified *way things are at a time t*, the way things go *thereafter* is *fixed* as a matter of *natural law*.[5]

Twentieth-century physics discovered that "the way things go" is *not* fixed as a matter of natural law.

The first fly in the ointment is called *the three-body problem*. Newton's laws had spectacular success predicting the motion of the planets around the sun. But the sun is much more massive than any of the planets and, hence, generates a deterministic system. When Newton's laws are applied to three or more celestial bodies of similar masses orbiting each other, no solution could be found. Laplace argued in the quote cited above that the future of the universe was determined according to Newton's laws: "[If we knew] all forces that set nature in motion...it would embrace in a single formula the movements of the greatest bodies of the universe and those of the tiniest atom." Yet, the "Laplace agenda"—showing that the universe was deterministic—proved to be obstinate. Even the case of three similar-sized orbiting bodies—the simplest case after two bodies—could not be solved. The situation was so central to nineteenth-century thought that Oscar II, King of Sweden, in 1887, offered a prize to anyone who could solve the three-body problem. The announcement required that the equations of the orbits "converges uniformly":

---

5   Hoefer, "Causal Determinism."

Given a system of arbitrarily many mass points that attract each according to Newton's law, under the assumption that no two points ever collide, try to find a representation of the coordinates of each point as a series in a variable that is some known function of time and for all of whose values the series converges uniformly.[6]

Henri Poincaré did not solve the problem but won the prize.[7] He showed that the three-body problem has no closed-form mathematical solution.[8] Poincaré invented the Poincaré map, a picture of how the orbits in the three-body problem (or other dynamic systems) vary from revolution to revolution. A Poincaré map of the three-body problem is shown in Figure 1.

The pattern of points x_T in the Poincaré map in Figure 1B are fractal, meaning that the complexity of this map grows without limit as the computation proceeds. No matter how accurately the orbits are modeled, or how many times the three bodies circle one another, the locations of the three bodies defy deterministic beliefs. Of course, the three bodies follow natural law—what else could they do—but natural law does not fix their positions in the future. The best that can be said about the positions of the three bodies is that they are more likely to be in certain areas than in others. The darker regions

6   Éric Charpentier et al., eds., *The Scientific Legacy of Poincaré* (American Mathematical Society, 2010), 165.

7   June Barrow-Green, *Poincaré and the Three Body Problem* (American Mathematical Society, 1996).

8   In 1772, the French mathematician Joseph-Louis Lagrange published five stable solutions to the three-body problem under the restriction that two of the bodies are much more massive than the third. Leonhard Euler had found three of these solutions earlier, but they are all called Lagrange points today. The James Webb Space Telescope is located at the second Sun-Earth Lagrange point. John Stillwell, *Mathematics and its History*, 3rd ed. (Springer, 2010), 254.

of the Poincaré map in Figure 1B are darker because the density of dots is higher in these regions than in the lighter regions.

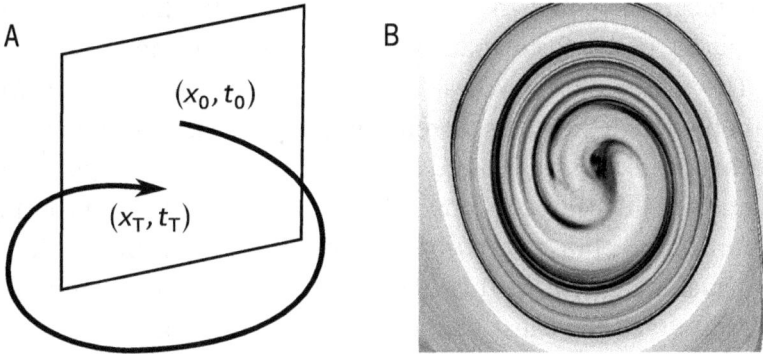

Figure 1A. The creation of a Poincaré map for the three-body problem. Each of the three bodies orbits in a three-dimensional loop as indicated for one body by the solid line. This orbit intersects the Poincaré plane at point x_0 at time $t = 0$ and at point x_T at time $t = T$. This provides two dots on the Poincaré plane. (1B) Repeating this process thousands of times with each of the three bodies generates thousands of points on the plane. The resulting set of dots provides the Poincaré map in 1B. Note that the dots plotted on the Poincaré map form a spiral pattern, but this pattern results from where the three bodies intersect the Poincaré plane and do not represent motion within the plane itself. The darker regions on a Poincaré map where the chaotic system spends significantly more time than the lighter regions are called *attractor states*.[9]

The determinist argument, "Sure, we can't predict the location of the planets in a three-body configuration, but the planets know where they need to go," falls on its face. Planets do not "know" anything. Existence does what it does; consciousness is just an observer.

---

9   Graphics reprinted with permission from Hassan Bevrani et al., *Grid Connected Converters: Modeling, Stability and Control* (Elsevier, 2022); and Rick Moeckel, "Chaos in the Three-Body Problem," Poincaré 100 Conference, November 2012, https://www-users.cse.umn.edu/~rmoeckel/presentations/PoincareTalk.pdf.

Only by putting consciousness in the driver's seat—making consciousness superior to existence—can one think that the orbits of three similar-sized bodies are determined.

Poincaré invented the Poincaré map, but he did not have the computing power to generate a map as detailed as the one in Figure 1B. The meteorologist Edward Lorenz revived chaos theory from near oblivion in the 1960s. Lorenz pioneered the use of computers in weather prediction and found that tiny changes in the inputs to the equations he developed produced enormous changes in the outputs. He coined the phrase *butterfly effect*, but this term is misleading. In the usual statement of this phenomenon, a butterfly flapping its wings in Brazil can cause enough difference in global weather simulation to alter the weather in New York City. The statement suggests that knowing every detail of the planet, even down to the level of butterflies, would allow us to predict the weather in New York City. This is not true. The butterfly is too big. One would have to zoom into smaller and smaller details, from the shape of the butterfly's antennae to the hairs on the antennae surfaces, to the pollen grains on the hairs, and the result would still be chaotic. No level of detail suffices to transform a chaotic system into a deterministic one. Existence is infinitely complex. Only an all-knowing deity could capture nature's unlimited complexity. In the real world, nature's complexity is beyond consciousness's reach.

Kerry Emanuel summarized the significance of Lorenz's work as:

By showing that certain deterministic systems have formal predictability limits, Ed put the last nail in the coffin of the Cartesian universe and fomented what some have called the third scientific revolution of the 20th century, following on the heels of relativity and quantum physics.[10]

---

10 "Edward Lorenz, Father of Chaos Theory and Butterfly Effect, Dies at 90,"...

While chaos theory represents "the third scientific revolution of the 20th century," this revolution is philosophical rather than practical. Relativity and quantum physics affect our daily lives—from atomic power and GPS to semiconductors and MRIs. The most common chaos-inspired product is a toy: a kinetic energy sculpture with two swinging arms based on the chaotic motion of the double pendulum.

People know quantum mechanics introduces uncertainty into the laws of physics. Fewer people know that randomness and uncertainty are there at the macroscopic level as well. Causal determinism requires predictability, and, in nature, predictability is seldom found. Existence exists and does its thing; mathematics can be used to model aspects of existence, but existence is too complex and unpredictable to be captured by theory. In the dispute of which is primary, consciousness or existence, existence has the upper hand.

THE ANSWER

Do we live in a clockwork universe? No, the universe is unpredictable. The Laplace agenda—showing that the future is fixed by the way things are today—is a chimera. The laws of physics don't converge to a single, unique outcome even in the simple case of three similar orbiting bodies. Determinism in physics, as in life, is a hollow concept.

...Massachusetts Institute of Technology News, April 16, 2008, https://news.mit.edu/2008/obit-lorenz-0416.

ARE ATOMS AND SUBATOMIC PARTICLES REAL?

LEVELS OF KNOWLEDGE

Our first level of certainty is the philosophic axioms, required by all knowledge and all thought.[1] The next level of certainty are ideas derived from sense perception: Stones are hard, water flows, air blows, and so on. Regardless of much information an individual learns later in life—rock quartz melts at 1,713°C, water turns into ice at 0°C, and air cooled to −194.35°C turns into a liquid—the child's initial concepts remain true in the context in which they were formed. Understanding the melting points of stones, ice,

---

1 Philosophic axioms are held implicitly by everyone, children and adults alike. To understand these axioms explicitly requires the process of thought.

and air does not alter the nature of stones, water, and air at ordinary room temperature and pressure.

At some point, the child learns that the solid stones of his or her youth are not so solid after all. Stones, and all matter, are composed of atoms, with spaces between them. The child has never seen an atom, but science teachers appear certain that the atomic theory is true.

This is a remarkable development, given that atoms were considered by many physicists to be hypothetical constructs at the beginning of the twentieth century rather than real objects.

In the fifth century BCE, the Greek philosopher Democritus argued, after his teacher Leucippus, that it is impossible to keep dividing matter infinitely many times and therefore matter must be made up of extremely tiny particles.[2] Democritus wrote that all objects are composed of many atoms united by random collisions, their forms and materials determined by the composition of atoms that make them up. Atoms float in a vacuum which Democritus called the *void*.

Democritus made a plausible argument, but his argument was philosophical rather than quantitative. Many doubted his hypothesis.[3]

The English chemist John Dalton published the first scientific evidence for atomism in 1804. Dalton observed that, if two elements form more than one compound, then the ratios of the masses of the second element that combined with a fixed mass of the first element, will always be ratios of small whole numbers. To explain this result, now called the law of multiple proportions, or sometimes Dalton's law, Dalton proposed that matter was composed of atoms, extremely small, indivisible particles identical in size and weight for each element.[4]

2   Bernard Pullman, *The Atom in the History of Human Thought*, trans. Axel Reisinger (Oxford University Press, 1998), 31–33.
3   Anthony Kenny, *Ancient Philosophy* (Oxford University Press, 2004), 26–28.
4   Mark I. Grossman, "John Dalton's 'Aha' Moment: The Origin of the Chemical...

Dalton takes us into the "possible" category, but far from certainty. A stronger argument for atomism was developed by the Scottish physicist James Clerk Maxwell.[5] Maxwell hypothesized in 1873 that gas is composed of "flying molecules":

> ...when "flying molecules" strike against a solid body in constant succession it causes what is called pressure of air and other gases.[6]

He showed mathematically that the pressure produced at a fixed temperature by these flying molecules is inversely propositional to the volume in which it is squeezed, that volume and Kelvin temperature are in direct proportion at constant pressure, and that the Kelvin temperature and the volume will be in direct proportion when the pressure is held constant. Using nothing except the idea that gas is a bunch of bouncing particles, Maxwell reproduced the ideal gas law.[7] The first part of this law is due to Robert Boyle, who noticed, in 1662, that the temperature of a gas is inversely propositional to its volume; the second part is due to Jacques Charles, who determined in 1802 that volume and Kelvin temperature are in direct proportion;[8] and

---

...Atomic Theory," *Ambix* 68, no. 1 (February 2021): 49–71, https://doi.org/10.108 0/00026980.2020.1868861.

5   Basil Mahon, *The Man Who Changed Everything: The Life of James Clerk Maxwell* (Wiley, 2003).

6   Gerald James Holton and Stephen G. Brush, *Physics, the Human Adventure: From Copernicus to Einstein and Beyond* (Rutgers University Press, 2001), 270.

7   The Austrian physicist Ludwig Boltzmann generalized Maxwell's results and formulated the Maxwell-Boltzmann statistical distribution of molecular speeds for ideal gases. Boltzmann went on to develop the concept of entropy and to formulate the second law of thermodynamics.

8   Jacques Charles was also a pioneer in flight. On December 1, 1783, Jacques Charles rose above Paris in the world's first hydrogen-filled balloon. Jean-François Pilâtre de Rozier beat Charles into the sky by rising in a hot-air balloon on October 15, 1783.

the third part is due to Gay-Lussac, who determined, in 1809, that pressure varies directly with the absolute temperature of the gas.

I recall being amazed as an undergraduate physics student by the power of math to derive the ideal gas law from first principles. The universe made sense to me; air was just a bunch of bouncing molecules.

Albert Einstein presented further evidence of the atomic theory of matter in a 1905 paper on Brownian motion.[9] Looking through a microscope, British botanist Robert Brown had observed a jittery motion of pollen grains suspended in water in 1827.[10] He had no explanation for the movement but ruled out life-related motion. Einstein hypothesized that Brownian motion was caused by water molecules jostling the Brownian particles. He evaluated the probability density of a particle moving a distance under these conditions. The result was a diffusion equation giving the density of Brownian particles throughout space. The quantitative predictions of this theory were verified experimentally by Jean Perrin in 1908. Perrin was awarded the Nobel Prize in Physics in 1926 "for his work on the discontinuous structure of matter."[11]

---

9   Albert Einstein, "Über die von der molekularkinetischen Theorie der Wärme geforderte Bewegung von in ruhenden Flüssigkeiten suspendierten Teilchen" [On the Movement of Small Particles Suspended in Stationary Liquids Required by the Molecular-Kinetic Theory of Heat], *Annalen der Physik* 322, no. 8 (1905): 549–60, https://doi.org/10.1002/andp.19053220806.

10   The Roman philosopher Lucretius described Brownian motion with remarkable acuity in approximately 60 BCE. In verses 113–140 of his epic poem *On the Nature of Things*, he uses dust particles viewed in sunbeams as a proof of the existence of atoms: "It originates with the atoms which move of themselves. Then those small compound bodies...are set in motion by the impact of [the atom's] invisible blows...so that those bodies are in motion that we see in sunbeams, moved by blows that remain invisible." While Lucretius was remarkably prescient in this description, dust particles are too large to exhibit Brownian motion. The movements Lucretius described are due to air currents.

11   Jean Baptiste Perrin, "Nobel Lecture," December 11, 1926, https://www.nobelprize.org/prizes/physics/1926/perrin/lecture/.

The atomic theory of matter became probable with the kinetic theories of gases and of Brownian motion, but not yet certain. What was needed was a way to "see" individual atoms, or at least to see individual atomic effects. Since Einstein's Brownian motion paper, thousands of procedures probing atoms have been developed. Scientists have even been able to image individual atoms using electron beans that impact upon crystalline structures. Figure 1 shows an image of the atomic structure of a praseodymium orthoscandate ($PrScO_3$) crystal, magnified 100 million times. The electron beam moves slowly across the sample, generating a speckle pattern in the detector as billions of electrons pass through or bounce around inside the sample before exiting. Computer software then calculates where the atoms are in the sample, creating the image.

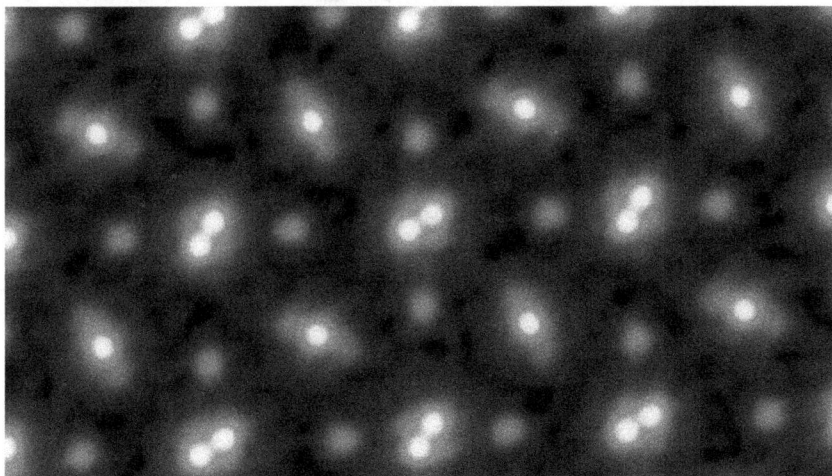

Figure 1. An electron ptychographic reconstruction of a praseodymium orthoscandate ($PrScO_3$) crystal, magnified 100 million times.[12]

---

12 Reproduced by permission from Zhen Chen et al., "Electron Ptychography Achieves Atomic-Resolution Limits Set by Lattice Vibrations," *Science* 372, no. 6544 (May 2021): 826–31, https://doi.org/10.1126/science.abg2533.

GOING DEEPER

No sooner was the concept of the atom
accepted in science than it fell apart.

Atoms were supposed to be the indestructible components of matter, but subatomic particles were encountered starting in the late nineteenth century. The first subatomic particle identified was the electron. Experiments by several physicists revealed rays were emitted when cathodes enclosed in a vacuum chamber were heated. In 1897, J. J. Thomson showed that cathode rays were particles, having both a charge e and a mass m, that their charge-to-mass ratio, $e/_m$, was independent of cathode material, and that the mass of the particles was less than one-thousandth of the mass of the least massive ion known: hydrogen.[13]

Henri Becquerel discovered radioactivity in Paris in 1896. Becquerel had noticed the blackening of a photographic plate wrapped in black paper after a uranium salt had been placed on it. He concluded that the uranium emitted an invisible form of radiation that could pass through the paper and cause the plate to react as if it had been exposed to light.

Ernest Rutherford's pursuit of Becquerel rays would forever change our concept of matter. In a series of experiments performed between 1898 and 1907 at McGill University in Montreal, Canada, Rutherford showed that radioactive decay represented a transmutation of elements, a discovery for which he received the Nobel

---

13   Thomson won the 1906 Nobel Prize in Physics. Eight of his research assistants (Charles Barkla, Niels Bohr, Max Born, William Bragg, Owen Richardson, Charles Wilson, Francis Aston, and Ernest Rutherford) also won Nobel Prizes. His son George Thomson won the Nobel Prize in Physics in 1937. J. J. Thomson, "Cathode Rays," *Philosophical Magazine* 44, no. 269 (October 1897): 293–316, https://doi.org/10.1080/14786449708621070.

Prize in Chemistry in 1908. He showed that different radioactive decay produced different types of radiation, each with a different power of penetration. He coined the terms *alpha rays*, *beta rays*, and *gamma rays* for the three most common radiation types. He showed that alpha particles were helium nuclei, and that radioactivity decreases exponentially with a half-life characteristic of the decay. In 1919, after he had moved to England, he bombarded nitrogen nuclei with alpha particles and discovered the emission of hydrogen nuclei. He named the emitted particle the *proton*.

In his most famous experiment, Rutherford bombarded a gold foil leaf with alpha particles. He had expected the particles to go straight through the thin foil. To his surprise, while most of the alpha particle went straight through, some particles were deflected back. In Rutherford's words:

> It was quite the most incredible event that has ever happened to me in my life. It was almost as incredible as if you fired a 15-inch shell at a piece of tissue paper and it came back and hit you.[14]

To explain why most of the particles go straight through, but some are deflected at large angles, or are even reflected, Rutherford proposed that most of the mass of an atom is concentrated in a small, dense nucleus, surrounded by a cloud of lighter orbiting electrons. Rutherford further hypothesized a new particle he called the *neutron*, with a mass roughly equal to the proton, but with zero charge, to explain why the mass of the nucleus is roughly twice that of the

---

14   Quoted in "Ernest Rutherford 1871–1937, New Zealand Physicist," *Oxford Essential Quotations*, 4th ed., ed. Susan Ratcliffe (Oxford University Press, 2016), https://www.oxfordreference.com/display/10.1093/acref/9780191826719.001.0001/q-oro-ed4-00009051.

number of protons in the nucleus.[15] In 1932, Frédéric and Irène Joliot-Curie detected a new type of radiation by bombarding beryllium with alpha particles.[16] Identifying the new particles as neutrons was left to Rutherford's associate James Chadwick, who was awarded the Nobel Prize in Physics for this work.[17]

The Danish physicist Niels Bohr extended Rutherford's atomic model by requiring electrons to maintain certain energy levels as they orbit the nucleus. In Bohr's model, an electron emits a photon of light when it transitions from a higher energy level to a lower level, as illustrated in Figure 2.

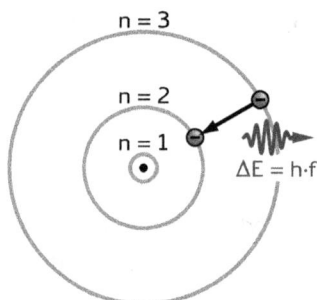

Figure 2. The Bohr model of a hydrogen atom with an electron transitioning from quantum level $n = 3$ to $n = 2$ and releasing a photon.

$$\Delta E = h \cdot f$$

15 Ernest Rutherford, "Bakerian Lecture: Nuclear Constitution of Atoms," *Proceedings of the Royal Society A: Mathematical, Physical and Engineering Sciences* 97, no. 686 (July 1920): 374–400, https://doi.org/10.1098/rspa.1920.0040.

16 Irène Curie was the daughter of Pierre and Marie Curie. Frédéric Joliot was an assistant of Marie Curie. They fell in love, married, and changed their surnames to Joliot-Curie in 1926. Frédéric and Irène Joliot-Curie jointly received the Nobel Prize in 1935. The only other married couple to receive a Nobel Prize was Irène's parents, Pierre and Marie Curie. The Curie family's five Nobels is the most to date by a single family.

17 James Chadwick, "The Existence of a Neutron," *Proceedings of the Royal Society A: Mathematical, Physical and Engineering Sciences* 136, no. 830 (June 1932): 692–708, https://doi.org/10.1098/rspa.1932.0112.

18 Image by MikeRun, derived from image by JabberWok (Wikimedia Commons, CC BY-SA 3.0).

THE ANSWER

Are atoms and subatomic particles real? Today, all the branches of the physical sciences are rooted in the atomic theory of matter. Atoms are real, as are subatomic particles, of that we can be certain. What is uncertain, and at odds with an objective view of the universe, is the Copenhagen interpretation of quantum mechanics, an interpretation that has consciousness driving physical events rather than existence. We explore this next.

DOES OBSERVING SOMETHING ALTER WHAT WE SEE?

WAVE/PARTICLE DUALITY

Contradictions do not exist, but early twentieth-century physicists said they had found one. This contradiction is illustrated by the double-slit experiment performed by Davisson and Germer in 1928. In this experiment, a beam of electrons is directed toward two slits cut into a solid membrane, followed by a screen that records where the electrons have landed.[1] As illustrated in Figure 1, observations reveal that the electrons passing through the slits form an interference pattern on the screen.

---

1   C. J. Davisson and L. H. Germer, "Reflection of Electrons by a Crystal of Nickel," *Proceedings of the National Academy of Sciences* 14, no. 4 (April 1928): 317–22, https://doi.org/10.1073/pnas.14.4.317.

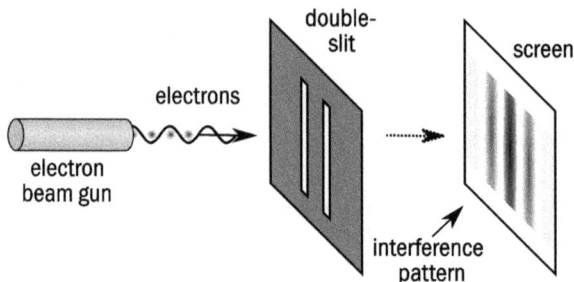

Figure 1. An interference pattern produced by an electron
beam passing through a double slit.[2]

The interference pattern from this experiment provides a conundrum. First, consider what would happen if the electrons were particles of ordinary matter such as bullets fired from a gun. With bullets, each bullet not hitting the solid membrane would go through one of the two slits, forming two distinct lines, one behind each slit. This is not what we see.

Now imagine a seawall with two entrances sheltering a harbor, analogous to the two slits in the double-slit experiment. Waves of water coming from the open sea would hit the seawall and enter through each opening. Each opening would form a new source of waves, with the waves from the openings spreading circularly outward in all directions. If the crest of a wave from one opening arrives at the same place as the equally deep trough from the other opening, the waves cancel each other out. If crest meets crest, they add to make a wave twice as tall. The series of crests and troughs form an *interference pattern* on the back wall of the harbor. This is the pattern that is observed in the double-slit experiment.

The mystery of the double-slit experiment is this: If a single electron is fired at the slits, the electron lands at a single location on the detector. One electron, one spot on the detector, clearly indicating

2   Image by Johannes Kalliauer (Wikimedia Commons, CC BY-SA 4.0).

particle-like behavior. However, as more electrons are fired at the slits, and more electron spot locations are recorded, a pattern emerges, the one illustrated in Figure 1. The electrons passing through the slits form an interference pattern, just like the one formed by water waves. Yet, each electron goes through the screen alone and arrives at the detector alone. How does one electron know how to interfere with the other electrons going through the screen before it or after it?

Richard Feynman was so puzzled by this experiment, that he said in his lectures on physics:

> We choose to examine a phenomenon which is impossible, absolutely impossible, to explain in any classical way, and which has in it the heart of quantum mechanics. In reality, it contains the only mystery. We cannot explain the mystery in the sense of "explaining" how it works. We will tell you how it works. In telling you how it works we will have told you about the basic peculiarities of all quantum mechanics.[3]

Notice that Feynman does not say, "I don't understand this result. I can't explain it." He says that it is "absolutely impossible" to explain it. He adds the caveat, "in any classical way." Feynman does not define what he means by this phrase, "in any classical way." My interpretation of this phrase is, "in a way consistent with logic and reason."

The basic equation of quantum mechanics, the equation that has caused such confusion, was published by Erwin Schrödinger in 1926.[4] To understand its origin, we need to go back to 1835, when

---

3   Richard Feynman et al., *The Feynman Lectures on Physics*, vol. 1 (California Institute of Technology, 1963), 37-1.

4   Erwin Schrödinger, "An Undulatory Theory of the Mechanics of Atoms and Molecules," *Physical Review* 28, no. 6 (December 1926): 1049–70, https://doi.org/10.1103/PhysRev.28.1049.

an Irish physicist named Sir William Hamilton developed what is called today *Hamiltonian mechanics*.[5] Suppose you throw a ball up in the air. To compute the trajectory of this ball using Newtonian mechanics, you compute the position and velocity of the ball, according to which velocity is the integral of the ball's acceleration and its position is the integral of its velocity. In this system, the gravitational constant gives the acceleration; the two variables computed are the velocity of the ball, v, and the height of the ball above the ground, h.

Hamilton introduced a new perspective on how to solve this problem. He said that instead of computing the velocity v and position h of the ball, one could compute its energy W and its momentum p. In classical mechanics, the two systems are equivalent: knowing the velocity and position of the ball at all times gives its energy and momentum at all times. Vice versa, knowing the energy and momentum of the ball at all times gives the ball's velocity and position.

The kinetic energy W_k of the ball is the energy the ball has because of motion. It is proportional to the square of its velocity v:

$$W_k = \tfrac{1}{2}\, m\, v^2$$

where m is the mass of the ball. The potential energy W_p of the ball is the energy it has due to being at a height h above the ground:

$$W_p = g\, m\, h$$

where g is the gravitational constant. The total energy of the system is the sum of the kinetic and potential energies:

$$H = W_k + W_p$$

where the quantity H is called the Hamiltonian of the system.

---

5   William Rowan Hamilton, "On a General Method in Dynamics," *Philosophical Transactions* (January 1834), https://doi.org/10.1098/rstl.1834.0017.

Schrödinger derived his equation by considering the energy and momentum of a quantum system. He evaluated the Hamiltonian of a quantum system and minimized it. The resulting equation, called the *Schrödinger equation*, is a wave equation that does not specify the exact path or location of an individual quantum particle. Instead, it provides a statistical distribution of quantum particle behavior. The values obtained from the Schrödinger equation are used to obtain the *probability density*, or the likelihood that a quantum particle will be at a specific location. It predicts the wavelike pattern observed by the detector in the double-slit experiment, but cannot predict where each individual electron will fall.

Quantum mechanics has been spectacularly successful. The Schrödinger equation has predicted the outcome of thousands of experiments. Extended in 1928 by Paul Dirac to include relativistic effects, the Dirac equation provides the fine details observed in the hydrogen spectrum.[6] The Dirac equation also predicts the existence of antimatter, a new class of particles with the same mass as ordinary matter but with the opposite charge. The positron, a positive version of the electron, was observed experimentally by Carl Anderson in 1932.[7] Further work by Wolfgang Pauli, Eugene Wigner, Hans Bethe, Richard Feynman, and others led to the development of quantum electrodynamics (QED).[8] With QED, the fine structure constant, a dimensionless number related to the spacing of the fine

---

6   Paul Adrien Maurice Dirac, "The Quantum Theory of the Electron," *Proceedings of the Royal Society A* 117, no. 778 (February 1928): 610–24, https://doi.org/10.1098/rspa.1928.0023.

7   Carl D. Anderson, "The Apparent Existence of Easily Deflectable Positives," *Science* 76 (1967): 238–39, https://doi.org/10.1126/science.76.1967.238.

8   Michael E. Peskin and Daniel V. Schroeder, *An Introduction to Quantum Field Theory* (Westview, 1995); and Richard Feynman, *QED: The Strange Theory of Light and Matter* (Princeton University Press, 1985).

lines appearing in the spectrum of the hydrogen atom, is computed with an accuracy within ten parts in one billion.[9]

Eight significant figures of accuracy between theory and measurement is a remarkable result. That is like measuring the width of the United States to the accuracy of a pebble. The issue with quantum mechanics is not its extraordinary scope or its extraordinary precision; it is with the interpretation of these results, an interpretation promulgated by twentieth-century physicists.[10] In Heisenberg's words:

> It is well known that the "reduction of the wave packets" always appears in the Copenhagen interpretation when the transition is completed from the possible to the actual. The probability function, which covered a wide range of possibilities, is suddenly reduced to a much narrower range by the fact that the experiment has led to a definite result, that actually a certain event has happened. In the formalism this reduction requires that the so-called interference of probabilities, which is the most characteristic phenomena of quantum theory, is destroyed by the partly undefinable and irreversible interactions of the system with the measuring apparatus and the rest of the world.[11]

9   G. Gabrielse et al., "New Determination of the Fine Structure Constant from the Electron g Value and QED," *Physical Review Letters* 97, 030802 (July 2006), https://doi.org/10.1103/PhysRevLett.97.030802.
10   Jagdish Mehra and Helmut Rechenberg, *The Historical Development of Quantum Theory*, vol. 4 (Springer, 1982), 266; and James Cushing, *Quantum Mechanics: Historical Contingency and the Copenhagen Hegemony* (University of Chicago Press, 1994).
11   Werner Heisenberg, "Criticism and Counterproposals to the Copenhagen Interpretation of Quantum Theory," in *Physics and Philosophy: The Revolution in Modern Science* (Harper, 1958), 114–28.

The "Copenhagen interpretation," as described by these words, has consciousness altering existence, not the other way around.

In Heisenberg's view, the quantum particle exists as a wave—a superposition of quantum states spread out over space—until someone measures it. Measurement, i.e., the observation by a consciousness of a quantum particle, collapses the wavefunction from its state as a wave into a point-like particle. "The universe is altered," Heisenberg says in effect, "by my observation." This is the opposite of objective knowledge. In objectivism, existence exists independently of consciousness, and consciousness is an observation of reality, not its creation.

Heisenberg's statement echoes Kant:

What we observe is not Nature itself, but Nature exposed to our method of questioning.[12]

John Stewart Bell, the originator of Bell's theorem, ridiculed the Copenhagen interpretation:

Was the wavefunction of the world waiting to jump for thousands of millions of years until a single-celled living creature appeared? Or did it have to wait a little longer, for a better qualified system... with a Ph.D.?[13]

The Copenhagen interpretation of quantum mechanics may be absurd, but many physics textbooks present it as if it was God's truth.

---

12   Marcelo Gleiser, *The Island of Knowledge: The Limits of Science and the Search for Meaning* (Basic Books, 2014).

13   John Stewart Bell, *Speakable and Unspeakable in Quantum Mechanics,* 2nd ed. (Cambridge University Press, 2004), 216.

THE ANSWER

Does observing something alter what we see? No, but advocates of the Copenhagen interpretation of quantum mechanics argue it does. This interpretation challenges the primacy of existence point of view of reality and makes the passive observation of atomic particles alter what the particles do. We show in answering *Question 23—Does an objective interpretation of quantum mechanics exist?* that an objective interpretation of quantum mechanics explains this philosophical conundrum.

DO WE LIVE IN A MULTIVERSE?

THE MULTIVERSE

n 1957, Hugh Everett had an idea. Everett had been struggling with the Copenhagen interpretation of quantum mechanics and found it to be wanting. The Copenhagen interpretation, as we have seen, is based on the idea that quantum mechanics is intrinsically indeterminate, with particles existing in a superposition of states until a measurement is made.[1] The superposition of states, the theory goes, collapses into a single determined state when it is observed. Everett argued that if the quantum state of a system is a superposition of two different quantum states, then both superposed states really exist.

---

1   Roland Omnes, "The Copenhagen Interpretation," in *Understanding Quantum Mechanics* (Princeton University Press, 1999), 41–54; and J. C. Polkinghorne, *The Quantum World* (Princeton University Press, 1986), 67.

Observation does not destroy one of them; it splits the universe into two branches.

The most famous example of this phenomena is Schrödinger's cat. In Schrödinger's words:

> A cat is penned up in a steel chamber, along with the following device (which must be secured against direct interference by the cat): in a Geiger counter, there is a tiny bit of radioactive substance, so small, that perhaps in the course of the hour one of the atoms decays, but also, with equal probability, perhaps none; if it happens, the counter tube discharges and through a relay releases a hammer that shatters a small flask of hydrocyanic acid. If one has left this entire system to itself for an hour, one would say that the cat still lives if meanwhile no atom has decayed. The first atomic decay would have poisoned it. The psi-function [wavefunction] of the entire system would express this by having in it the living and dead cat (pardon the expression) mixed or smeared out in equal parts.[2]

According to the Copenhagen interpretation, the cat is in a super-position of states, both dead and alive. Opening the box collapses the states; one finds either a dead cat or a live one. Everett found Schrödinger's "living and dead cat...smeared out in equal parts" to be nonsensical. His solution? The quantum world branches into different worlds, Everett said, one world with a live cat, another world with a dead cat.

---

2   Erwin Schrödinger, "Die gegenwärtige Situation in der Quantenmechanik," *Naturwissenschaften* 23, no. 48 (November 1935): 807–812; and John D. Trimmer, "The Present Situation in Quantum Mechanics: A Translation of Schrödinger's 'Cat Paradox' Paper," *Proceedings of the American Philosophical Society* 124, no. 5 (October 1980): 323–38, https://www.jstor.org/stable/986572.

Schrödinger had meant his thought experiment to show the absurdity of the Copenhagen interpretation of quantum mechanics. Instead, it prompted Everett to develop the multi-worlds interpretation (MWI) of quantum mechanics. MWI has become a leading interpretation of the ultimate nature of reality.[3] Max Tegmark has enumerated four levels of other worlds beyond the one we see.[4] Brian Greene presents nine types in his taxonomy of other universes.[5] A poll of seventy-two quantum field theorists showed a 58 percent agreement with "Yes, I think MWI is true."[6] In the "yes" column were Stephen Hawking, Richard Feynman, and Murray Gell-Mann.

The multi-worlds interpretation (MWI) of quantum mechanics is absurd on its face. In MWI, every quantum event generates another world. You exist not only as yourself but as innumerable different copies, one for each branch in MWI. Further, there is no way to know

---

3   Bryce S. DeWitt, "Quantum Mechanics and reality," *Physics Today* 23, no. 9 (September 1970): 30–5; Max Tegmark, "The Interpretation of Quantum Mechanics: Many Worlds or Many Words?," *Fortschritte der Physik* 46, no. 6–8 (April 1999): 855–62, https://doi.org/10.1002/(SICI)1521-3978(199811)46:6/8<855::AID-PROP855>3.0.CO;2-Q; Zeeya Merali, "This Twist on Schrödinger's Cat Paradox Has Major Implications for Quantum Theory," *Scientific American*, August 17, 2020, https://www.scientificamerican.com/article/this-twist-on-schroedingers-cat-paradox-has-major-implications-for-quantum-theory/; George Musser, "Quantum Paradox Points to Shaky Foundations of Reality," *Science*, August 17, 2020, https://www.science.org/content/article/quantum-paradox-points-shaky-foundations-reality; and Simon Saunders et al., eds., *Many Worlds? Everett, Quantum Theory and Reality* (Oxford University Press, 2010).

4   Max Tegmark, "Parallel Universes," *Scientific American* 288, no. 5 (May 2003): 40–51, https://doi.org/10.1038/scientificamerican052003-31q16fbiXWSSwJMsKfi51X.

5   Brian Greene, *The Hidden Reality: Parallel Universes and the Deep Laws of the Cosmos* (Knopf, 2011).

6   Frank Tipler, *The Physics of Immortality: Modern Cosmology, God and the Resurrection of the Dead* (Doubleday, 1994), 170–71.

if the other worlds exist; communication with the other MWI worlds is not possible in principle. Since we cannot prove their existence, we must accept MWI on faith.

Let us contemplate this. After centuries of progress, centuries that saw men use physics to put men on the moon and cell phones in men's pockets, several leading physicists today hold the position that multiple universes exist, a theory that can be accepted only on faith.

The multiverse[7] contemplated by physicists is not science fiction, even though conceptualizing a universe different from our own is a common literary device, useful for entertaining and educating the human soul. *Atlas Shrugged*, Ayn Rand's magnum opus, is set in an alternate universe. Every work of fiction takes place in a world that does not exist. MWI physicists, on the other hand, hold that the multiverse exists. Stephen Hawking said that MWI is "trivially true."[8] Nobel laureate Murray Gell-Mann wrote, "...it is unnecessary to become queasy conceiving of many parallel universes, all equally real."[9]

Skeptics focus on the impossibility of empirical testability and falsifiability of the multiverse hypothesis. George Ellis and Joe Silk argue that attempts to exempt speculative theories of the universe from experimental verification undermine science.[10] Ellis writes:

---

7   MWI is just one example of a hypothetical multiverse. Speculations of multiple universes predate Socrates, include such philosophers as William James, and encompass imagined universes formed alongside ours when the Big Bang happened. See New Scientist, *The Universe Next Door: A Journey Through 55 Alternative Realities, Parallel Worlds and Possible Futures* (Nicholas Brealey Publishing, 2017), 12.

8   Martin Garner, *Are Universes Thicker than Blackberries? Discourses on Godel, Magic Hexagrams, Little Red Riding Hood, and Other Mathematical and Pseudoscientific Topics* (W.W. Norton, 2003), 10.

9   Murray Gell-Mann, *The Quark and the Jaguar: Adventures in the Simple and the Complex* (Owl Books, 1994), 138.

10   George Ellis and Joe Silk, "Scientific Method: Defend the Integrity of Physics," *Nature* 516, (December 2014): 321–23, https://doi.org/10.1038/516321a.

As skeptical as I am, I think the contemplation of the multiverse is an excellent opportunity to reflect on the nature of science and on the ultimate nature of existence: why we are here... In looking at this concept, we need an open mind, though not too open. It is a delicate path to tread. Parallel universes may or may not exist; the case is unproved. We are going to have to live with that uncertainty. Nothing is wrong with scientifically based philosophical speculation, which is what multiverse proposals are. But we should name it for what it is.[11]

An individual's confidence in his or her mind is a consequence of knowing what is true and what is false. Knowledge—the totality of all correct statements—requires standing on a firm epistemological foundation. "Parallel universes may or may not exist; the case is unproved" and "[there is] nothing wrong with scientifically based philosophical speculation" undercuts this foundation and makes us less sure of ourselves. These statements are profoundly anti-mind.

NOT EVEN WRONG

Scientifically based philosophical speculation has done actual damage to science and philosophy. The speculation "Quantum particles exist in a superposition of states that collapses when observed" has turned some physicists, such as Richard Feynman, into witch doctors chanting incantations they say they do not understand. Speculations such as Schrödinger's cat and the multiverse drive rational men away from physics and let charlatans in.

---

11   George F. R. Ellis, "Does the Multiverse Really Exist?," *Scientific American* 305 no. 2 (August 2011), 38–43, https://www.scientificamerican.com/article/does-the-multiverse-really-exist/.

Epistemology is the science of rational thinking, the rules to follow to ensure consistency between thoughts and existents. These rules begin with definitions. The rules continue with logic, Aristotle's laws that prevent contradictions. Above all, to be logical, a person needs to identify things that exist.

Objectivism holds that logic is the art of noncontradictory identification.[12]

Young children are consumed by perception, reality, and imagination. Play is the way to learn what is real, what is not, what is achievable, and what is only imagined. A child's concepts are easily resolved by pointing at existents. Flying reindeer and Santa Claus give way to building airplanes and producing wealth as we grow up. At some point, concepts become too complicated to be resolved by pointing, and a person must rely on definitions to maintain his or her link to reality.

A concept substitutes one symbol (one word) for the enormity of the perceptual aggregate of the concretes it subsumes.[13]

Definitions provide the stepping stones by which man integrates concepts into a systematic whole. As described in *Question 10—Do mathematical objects exist?*, a proper definition provides a one-to-one correspondence between concepts and existents.[14] We make nature's complexity manageable by identifying existents through their essential characteristics and labeling each existent with a word. The process of reason is unforgiving. Using a word in a sentence

---

12   FNI, 125.
13   ITOE, 64.
14   ITOE, 40.

without knowing what it means, or defining a word through nonessential characteristics, severs links in the logical chain.

"Parallel universes may or may not exist; the case is unproved," is a particularly insidious statement. Perceptual observation is automatic. We all see the same forest and trees. If any mistakes occur, they happen "in the process of reason." Is the forest haunted? Is there a ghost behind the tree? A person who "sees" things that are not there is considered to be delusional, but a person who declares, "Ghosts may or may not exist; the case is unproved" damages cognition. Ideas that cannot be traced back to sense perception are "floating abstractions," Ayn Rand's term for concepts detached from existents.[15]

A concept is not a product of arbitrary choice, but results from focusing man's mind on that which exists.

A definition is the condensation of a vast body of observations—and stands or falls with the truth or falsehood of these observations.[16]

Existence simply is. Man must determine what is. A concept is said to be true if it correctly describes that which exists; false if the concept does not correspond to that which exists. Science is the grasp of nature, not its creation. Knowledge increases step by step, as observations and thought add depth to man's understanding of existents. "All thought, argument, proof, refutation must start with that which exists."[17]

---

15   Abstractions made in the service of fiction are not "floating abstraction." They serve a valid cognitive purpose as discussed in Question 40—What's the purpose of art?

16   ITOE, 48.

17   OPAR, 168.

A proposition built from concepts attached to reality can be true or false, depending on the consistency of the argument. According to Leonard Peikoff:

> Proof is the process of establishing a conclusion by identifying the proper hierarchy of premises. In proving a conclusion, one traces backward the order of logical dependence, terminating with the perceptually given.[18]

For example, if one wants to show that electrons exhibit wave-like behavior, one begins with: What is an electron? How does one accelerate an electron? What is a double slit? What pattern of electrons would appear on the detector if the electron behaved in a wavelike manner? Each one of the answers to these questions would be a proposition that needs further examination until each item is directly perceived by the senses. Of course, one does not need to perform this lengthy process when everything is as expected. But if an experiment does not produce the expected result, the scientist traces every step in the experiment back to its source until an error in the experiment, or an error in thinking, is located.

Here, the proposition, "Electrons exhibit wavelike behavior," is true. The interference pattern appears on the detector screen, a pattern characteristic of waves. Had the proposition been "Electrons do not exhibit wavelike behavior," and one observed an interference pattern on the screen, this proposition would be false. In either case, we label the proposition true or false by comparing it to the observation of reality.

Suppose, however, one makes a proposition untied to reality. "There is a ghost in the forest." This proposition is not true; there is nothing in the forest at which to point and say, "This is a ghost." But

---

18   OPAR, 138.

it is not false either, at least in the sense there is nothing you can point at and say, "Your idea of a ghost is not true because it doesn't match sense data." There is no sense data with which to compare a ghost. You can say, "I don't see the ghost," but the believer will say, "Just because you don't see the ghost doesn't mean it doesn't exist." It is not possible to "prove" there are no ghosts.

A "floating abstraction" is *arbitrary* because it is untied to reality. One can make up any idea by combining concepts in an arbitrary fashion: A ghost is the spirit of a dead man walking around on earth; God is a brain-free consciousness creating material things out of nothing; a parallel universe is a universe just like ours but somehow apart from everything that exists.

The arbitrary is the enemy of sanity. Man must follow the process of reason to achieve efficacy. As explained in *Question 8—Does induction prove anything?*, there is only one universe in which everything is interconnected.[19] It follows from this that all knowledge is interconnected. Certainty and confidence follow from an integrated view of the world.

Knowledge is the totality of all correct thoughts.

Arbitrary statements destroy this integration. Disconnected from existents, both perceptually and conceptually, the arbitrary splits man's mind from that which exists. Uncertainty and doubt follow. How do I know there are no ghosts, or God, or other universes? Isn't it better to be an agnostic in all things? Leonard Peikoff writes:

Philosophically, the arbitrary is worse than the false. The false has a relation, albeit negative, to the facts of reality; it has reached the field of human cognition and invoked its methods, even though

---

19 ITOE, 39.

an error has been committed in the process. This is radically different from the capricious. The false does not destroy a man's ability to know; it does not nullify his grasp of objectivity; it leaves him the means of discovering and correcting his error. The arbitrary, however, if a man indulges in it, assaults his cognitive faculty; it wipes out or makes impossible in his mind the concept of rational cognition and thus entrenches his inner chaos for life.[20]

Since there is no way to bring arbitrary statements into the realm of reason, they must be rejected. An arbitrary claim invalidates itself. Nothing is gained by contemplating ghosts except fear and uncertainty; nothing is gained by worshiping God except paranoia and superstition; nothing is gained by theorizing an alternate universe unconnected to existence.

The multiverse is arbitrary. It is not even wrong.

THE ANSWER

Do we live in a multiverse? No, there is only one universe. The multiverse is not observable in principle and depends entirely on the speculative Copenhagen interpretation of quantum mechanics. We must base scientific theories on experimental observations and avoid speculating about things that can never be observed.

---

20   ITOE, 166.

DOES AN OBJECTIVE INTERPRETATION OF QUANTUM MECHANICS EXIST?

BOHMIAN MECHANICS

The absurdity of the Copenhagen interpretation becomes even stranger when we discover that Louis de Broglie developed a logical, consistent, and realism-based interpretation of quantum mechanics the 1920s but was persuaded to abandon his theory because he was "discouraged by criticisms which [it] roused."[1] Rediscovered by David Bohm in 1952, the de Broglie–Bohmian approach to quantum mechanics postulates that particles have

---

1   Louis de Broglie, foreword to *Causality and Chance in Modern Physics*, by David Bohm (Routledge, 1957), x.

well-defined positions at all times but are guided through space by *pilot waves*.[2] In the de Broglie–Bohm theory, a single wavefunction exists in the entire universe, the sum of all the wavefunctions of all the particles that exist. This wavefunction satisfies the Schrödinger equation as before, but is augmented by a *guiding equation* for each particle. The guiding equation provides the velocity and position of the particle along its trajectory in space.[3] The analogy is to a swarm of celestial bodies moving in response to their common gravitational field, except that a swarm of quantum particles rides the pilot waves of the universal wavefunction.

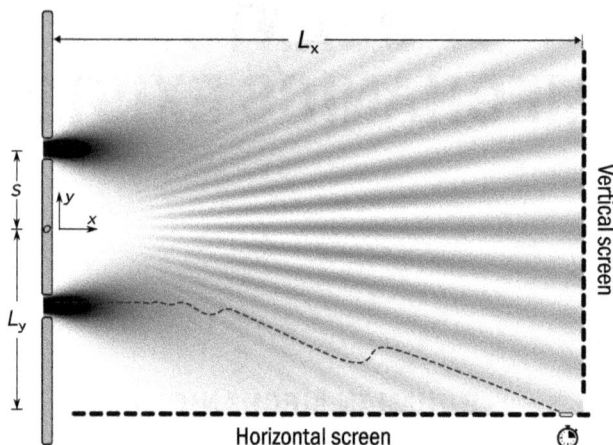

Figure 1. The trajectories of electrons passing through a double slit computed using Bohmian mechanics. The dashed black line shows a typical Bohmian trajectory—with recursive movements—that arrives at the horizontal screen.[4]

2   David Bohm, "A Suggested Interpretation of Quantum Theory in Terms of 'Hidden' Variables," *Physical Review* 85 (January 1952): 166–92, https://doi.org/10.1103/PhysRev.85.166.

3   Sheldon Goldstein, "Bohmian Mechanics," *The Stanford Encyclopedia of Philosophy*, summer 2024 ed., eds. Edward N. Zalta and Uri Nodelman, https://plato.stanford.edu/archives/sum2024/entries/qm-bohm/.

4   Reproduced by permission from Ali Ayatollah Rafsanjani et al., "Can the...

The Bohmian trajectories of electrons going through the double-slit experiment is presented in Figure 1. While each trajectory passes through only one slit, the wave passes through both. The electrons follow different trajectories because the wavefunctions of the electrons are entangled. The common wavefunction causes more Bohmian trajectories to end at the points on the screen where the squared amplitude of the wavefunction is large, and few to end where the squared amplitude is small. The interference pattern seen on the screen is a result of this common wavefunction.

In Bohmian mechanics, the particles are the primary elements of reality while the wavefunction is secondary or derivative, analogous to mass as primary and gravity as derivative in Newtonian mechanics, and charge as primary and electric field as derivative in electromagnetics. Particles act according to physical laws with no need for an observer. To make this point fully real, in 2006, Yves Couder and Emmanuel Fort created a macroscopic analog of the de Broglie–Bohm system. They showed that small drops of a viscous fluid bounce off a bath of the same fluid if the bath is made to oscillate vertically. The bouncing drop creates a wave on the surface of the liquid. Couder and Fort demonstrated that this liquid drop/wave combination is analogous to the quantum particle/pilot-wave system. They were able to duplicate the double-slit experiment using this hydrodynamical system. Since then, several of the quantum phenomena have been replicated on the macroscopic level including tunneling and quantized orbits.[5] Richard Feynman's assertion that, "[It] is impossible, absolutely impossible, to explain [the double-slit

...Double-Slit Experiment Distinguish Between Quantum Interpretations?," *Communications Physics* 6 (July 2023): 195, https://doi.org/10.1038/s42005-023-01315-9.

5   John W. M. Bush, "The New Wave of Pilot-Wave Theory," *Physics Today* 68, no. 8 (2015): 47–53, https://doi.org/10.1063/PT.3.2882.

experiment] in any classical way" has been refuted twice, once theoretically through the pilot-wave prescription and a second time by observing drops of bouncing fluids.[6]

Of these two interpretations of quantum mechanics, one claims that an electron or other quantum particle is in a superposition of wavefunction states until it is observed, at which point the wavefunction collapses to give the electron a definite position. The other interpretation says that the particle exists in a definite location all the time but is guided by a pilot wave. The first solves the particle-wave duality of the electron by postulating multiple smeared-out states of the electron as it travels through a slit, but these wavefunctions collapse into a single entity when the particle is observed. The second provides particle-wave duality by saying that the particle always exists as a particle but is guided by the wavefunction. In both cases, the mathematical recipe for computing the wavefunction is the same: The wavefunction satisfies Schrödinger's equation in both cases. Both interpretations give identical results.

Maudlin writes:

A physical theory should contain a physical ontology: What the theory postulates to exist as physically real. And it should also contain dynamics: laws (either deterministic or probabilistic) describing how these physically real entities behave. In a precise physical theory, both the ontology and the dynamics are represented in sharp mathematical terms... A precisely defined physical theory, in this sense, would never use terms like "observation," "measurement," "system," or "apparatus" in its fundamental postulates. It would instead say precisely what exists and how it behaves.[7]

---

6   Feynman et al., *The Feynman Lectures on Physics*, 37-1.
7   Maudlin, *Philosophy of Physics*, 4–5.

Of the two leading interpretations of quantum mechanics, only Bohmian mechanics satisfies Maudlin's ontological criteria.

Bohmian mechanics says precisely what exists and how it behaves. The Copenhagen interpretation of quantum mechanics mixes measurement and particle behavior into one mysterious sum. Yet, the Copenhagen interpretation dominates the world of physics today. The mathematical physicist Sheldon Goldstein said of Bohm's theory in 2016:

> There was a time when you couldn't even talk about it because it was heretical. It probably still is the kiss of death for a physics career to be actually working on Bohm, but maybe that's changing.[8]

Change is in the air. Two groups of theoretical physicists have proposed experiments that distinguish between the two leading interpretations of quantum mechanics.[9] Bohmian mechanics provides time-of-flight data for the particles going through the double slit; the Copenhagen interpretation does not.[10] The stakes could not be higher: The results will determine whether observation does indeed alter the behavior of quantum particles, or if these particles follow consciousness-free laws.

------

8   Anil Ananthaswamy, "Quantum Weirdness May Hide an Orderly Reality After All," New Scientist, February 19, 2016, https://www.newscientist.com/article/2078251-quantum-weirdness-may-hide-an-orderly-reality-after-all/.
9   Ayatollah Rafsanjani et al., "Can the Double-Slit Experiment Distinguish Between Quantum Interpretations?"; and Dirk-André Deckert Siddhant Das et al., "Double-Slit Experiment Revisited," arXiv (November 2022), https://doi.org/10.48550/arXiv.2211.13362.
10   Anil Ananthaswamy, "Can We Gauge Quantum Time of Flight?," Scientific American, October 21, 2021, https://www.scientificamerican.com/article/this-simple-experiment-could-challenge-standard-quantum-theory/.

EXPERIMENTAL METAPHYSICS

In 1935, Albert Einstein, along with coauthors Boris Podolsky and Nathan Rosen, spotted a problem with quantum mechanics. They noticed that certain quantum effects could travel faster than the speed of light. Einstein had discovered the universal speed limit—nothing can outpace light—in 1905. And here, a mere thirty years later, quantum particles were said to violate it.

The problem arose with *spin*, a quantum quantity with no classical counterpart. In classical mechanics, the three-component vector defining the angular momentum of a spinning ball provides a clear and unambiguous expression. The ratios of the values of these three components give the orientation of the axis of the spinning ball and the magnitude of the vector gives the magnitude of the angular momentum. In quantum mechanics, however, angular momentum is quantized, meaning it can take on only discrete values.[11] Vector math does not work for spin—if you measure one component of angular momentum for a quantum particle, the other values cannot satisfy the quantization conditions these components need to take.

Einstein, Podolsky, and Rosen (EPR) considered a thought experiment in which a pair of entangled particles are emitted from a source in opposite directions. *Entangled particles* share a single, joint wavefunction. In David Bohm's interpretation of the EPR experiment, two *spin meters* are set up, one to the left (L) of the source, the other to the right (R), as indicated in Figure 2.[12] Measurements of each particle's

---

11   The values of spin for protons and other hadrons are $\pm\hbar$ while electrons and other leptons have spin $\pm\hbar/2$ where $\hbar = h/(2\pi)$ and h is Planck's constant. Planck's constant is the fundamental unit in quantum mechanics. The energy of a photon is Planck's constant times the photon's frequency. The energy of a matter wave is Planck's constant times its wavelength divided by the associated particle momentum.

12   David Bohm, "A Suggested Interpretation of the Quantum Theory in Terms...

spin components are made some distances away from the source, the orientations for measuring the spin chosen independently by the different individuals observing the left and right apparatuses.

Consider the case of a pair of particles, one particle with spin up (+1 spin) and the other particle with spin down (−1 spin), entangled to form a single wavefunction having zero spin. Einstein, Podolsky, and Rosen pointed out that measuring the spin at one apparatus would immediately collapse the wavefunction and reveal the spin of the particle at the other apparatus. For example, if the spin measured at the left apparatus is +1, then the spin of the entangled particle at the right apparatus must be −1. This effect is immediate. Einstein called this instantaneous response between two widely separated particles "spooky action at a distance."[13]

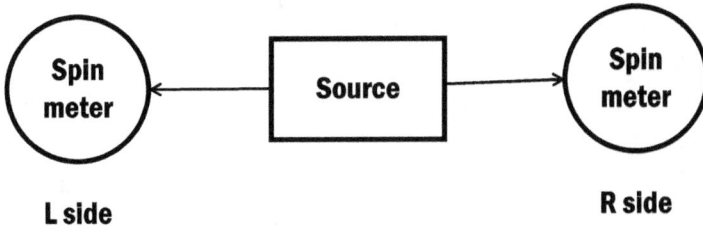

Figure 2. A schematic illustration of the EPR experiment. Particle pairs with entangled spin are emitted in opposite directions and components of these spins measured after the particle have traveled far from the source.

In the Copenhagen interpretation of quantum mechanics, the particles do not have any definite spin until they are measured. The first spin measurement determines the particles' spin, and the result

---

...of 'Hidden' Variables," *Physical Review* 85, no. 166 (January 15, 1952), https://journals.aps.org/pr/abstract/10.1103/PhysRev.85.166.

13   Albert Einstein et al., *The Born-Einstein letters: Correspondence Between Albert Einstein and Max and Hedwig Born, 1916–1955* (Macmillan, 1971), 158.

is purely random. For instance, if the first measurement is an x-axis spin measurement on the L-particle, the L-particle will spin either clockwise or counterclockwise about the x-axis with equal probability. This measurement also instantaneously influences the spin of the distant R-particle. If the L-particle spins clockwise (counterclockwise) around the x-axis, the R-particle will simultaneously spin in the opposite direction.

Einstein, Podolsky, and Rosen's thought experiment inspired John Stewart Bell to prove a limit on the correlations of outcomes in far-apart measurements. He showed that the Copenhagen interpretation of quantum mechanics predicted stronger statistical correlations than was possible if quantum effects were local.[14] Tests of Bell's theorem are called *experimental metaphysics* because Bell's theorem isn't about a particular physical theory. Instead, Bell's theorem is a general proof that reveals possible constraints on metaphysical assumptions.[15]

Numerous experiments have shown since Bell proved his theorem that quantum mechanics violates the metaphysical assumptions of locality and realism.[16] John Clauser, Alain Aspect, and Anton Zeilinger received the 2022 Noble Prize as three pioneers in this field.

The two leading proposals to explain the faster-than-light correlation of quantum phenomena are:

---

14 J. S. Bell, "On the Einstein Podolsky Rosen Paradox," *Physics Physique Fizika* 1, no. 3 (November 1964): 195–200, https://doi.org/10.1103/ PhysicsPhysiqueFizika.1.195.

15 Amanda Gefter, "'Metaphysical Experiments' Probe Our Hidden Assumptions About Reality," *Quanta Magazine*, July 30, 2024, https://www.quantamagazine. org/metaphysical-experiments-test-hidden-assumptions-about-reality-20240730/?mc_cid=48655a0431&mc_eid=e4f2d386ea.

16 Wayne Myrvold et al., "Bell's Theorem," *The Stanford Encyclopedia of Philosophy*, spring 2024 ed., eds. Edward N. Zalta and Uri Nodelman, https:// plato.stanford.edu/archives/spr2024/entries/bell-theorem/.

1. **The Copenhagen Interpretation**. Particles do not exist prior to observation, only a superposition of wavefunctions exists providing all the possible states a potential particle could take. As if by magic, particles pop into view when an experiment is performed. If two particles are entangled, observation of one of the pair of particles instantaneously collapses the other particles' superposition of wavefunctions regardless of the distance between the particles. This provides "spooky action at a distance," in Einstein's words.

2. **The de Broglie–Bohm Interpretation**. Particles always exist in defined locations and possess defined states. Each particle contributes to a universal wavefunction that permeates all space. In this case, observation is prosaic. It simply reveals the state of the particles the way they are. Results of measurements of two entangled particles close together are the same as measurements made when the particles are far apart. This is because a particle's properties are attributes of the particle, not of the wavefunction.

John Stewart Bell wrote in 1989:

De Broglie showed in detail how the motion of a particle, passing through just one of two holes in screen, could be influenced by waves propagating through both holes. And so influenced that the particle does not go where the waves cancel out but is attracted to where they cooperate. This idea seems to me so natural and simple, to resolve the wave-particle dilemma in such a clear and ordinary way, that it is a great mystery to me that it was so generally ignored.[17]

---

17   Bell, *Speakable and Unspeakable in Quantum Mechanics*, 191.

The mystery remains: Why are people so keen to assign consciousness an active role in quantum mechanics when the reality-based approach is so natural and simple?

THE ANSWER

Does an objective interpretation of quantum mechanics exist? Yes. A theory developed by Louis de Broglie in the 1920s and resurrected by David Bohm in the 1950s provides an objective theory of quantum mechanics. In Bohmian mechanics, quantum particles exist in defined positions at all times and are guided along trajectories by pilot waves. The Copenhagen interpretation of quantum mechanics and the de Broglie–Bohm theory provide identical results, but the Copenhagen interpretation requires instantaneous "action at a distance" between two entangled but widely separated particles. Bohmian mechanics require no such "spooky" behavior.

ARE THE LAWS OF THE UNIVERSE SYMMETRIC?

NATURE'S SYMMETRY

Emmy Noether could not find a job. Ranked "the most important woman in mathematics" by Albert Einstein in 1935, Noether was not allowed to teach in her early professional career because she was a woman. She worked at the Mathematical Institute of Erlangen for seven years without pay after completing her doctorate.[1] In 1915, she was invited by the world-renowned mathematician David Hilbert to join the mathematics department at the University of Göttingen, a major hub of mathematical research. However, she

---

1   Albert Einstein, "The Late Emmy Noether: Professor Einstein Writes in Appreciation of a Fellow-Mathematician," *New York Times*, May 4, 1935, https://www.nytimes.com/1935/05/04/archives/the-late-emmy-noether-professor-einstein-writes-in-appreciation-of.html.

was not allowed to lecture until 1919. She was expelled from the University of Göttingen in 1933 when the German Nazi government dismissed all Jews from university positions. After her dismissal, Noether moved to Bryn Mawr College, in Philadelphia, Pennsylvania, USA, where she lectured for two years until her death.[2]

Much of physics today is based on Noether's work. She showed that a symmetry embedded in a physical law implies the conservation of the corresponding physical quantity, and vice versa.[3] For example, if a physical system behaves the same regardless of how it is rotated in space, Noether's theorem states that the angular momentum of the system is conserved. The physical system itself need not be symmetric; a tumbling space station conserves angular momentum despite its asymmetry. It is the laws of its motion that are symmetric.

Symmetry allows scientists to study physical systems in entirely new ways. For example, Johannes Kepler's elliptical orbits are mathematically equivalent to particles moving freely on the surface of a hypersphere of four dimensions.[4] The physical system, a planet orbiting a sun, is easier to solve abstractly by using four variables in "hyperspace," rather than directly in terms of the three physical coordinates of ordinary space.[5] Symmetry is imbedded in the physics: The orbits of planets around the sun are symmetric under rotations in four-dimensional "function space."

---

2   Emily Conover, "In Her Short Life, Mathematician Emmy Noether Changed the Face of Physics," *ScienceNews*, June 12, 2018, https://www.sciencenews.org/article/emmy-noether-theorem-legacy-physics-math.

3   Jozef Hanca et al., "Symmetries and Conservation Laws: Consequences of Noether's Theorem," *American Journal of Physics* 72, no. 4 (April 2004): 428–35, https://doi.org/10.1119/1.1591764.

4   William R. Hamilton, "A New Method of Expressing in Symbolic Language the Newtonian Law of Attraction," *Proceedings of the Royal Irish Academy* 3 (1844): 344–53, https://www.jstor.org/stable/20489607.

5   Herbert Goldstein, *Classical Mechanics*, 2nd ed. (Addison Wesley, 1980), 102–105, 421–22.

Twentieth-century physics took symmetry to heart. Nature, to be understood, physicists said, must be examined in terms of its *symmetries*. Symmetry, in this sense, is not a property of an object, but a fundamental aspect of existence.

Symmetries in physical laws shine a light on existence.

We can easily visualize symmetry in ordinary space. Consider the hexagon presented in Figure 1. We can rotate this figure counterclockwise by 60° and obtain exactly the same figure. Let R be the 60° rotation operator. The points z_i, $i = 0,..., 5$ satisfy the relation:

$$z_{i+1} = Rz_i, i = 0, ..., 5 \text{ with } z_6 \rightarrow z_0$$

The six points z_i, $i = 0,..., 5$ form a *group*. No matter how many times you apply the operator R to a point z_i, you will obtain a point z_i in the group. The key takeaway from group theory is this seemingly obvious point: *A hexagon is invariant under a 60° rotation.*

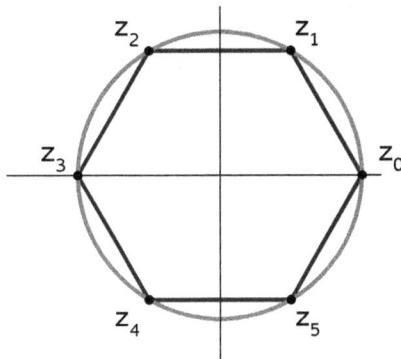

Figure 1. A hexagon has six-fold rotational symmetry. Each of the points z_i lands on another point in the group z_i, $i = 0,..., 5$, through a 60° rotation.[6]

---

6   Image by Jakob.scholbach (Wikimedia Commons, CC BY-SA 3.0).

The analog between vectors and functions is explained by considering two vectors: A and B, as shown in Figure 2. In ordinary space,
the projection of vector A onto vector B is defined as the component
of A pointed in the same direction as B. This projection is what your
eye sees when someone spins an arrow in front of you. The apparent
length of the arrow ranges from its actual length, written as |A| when
the arrow is broadside to you, to zero when the arrow is pointed
either toward you or away from you. Trigonometry gives the appar-
ent length, or the projection, of A onto B as $\cos \theta\, |A|$ where θ is the
angle between vectors A and B. Writing A and B in terms of their x
and y components, a_x and a_y for A, b_x and b_y for B, we evaluate:

$$A \bullet B = a_x\, b_x + a_y\, b_y$$

The quantity $A \bullet B$ is called the dot product of the vectors A and B.
After some algebra, we obtain:

$$A \bullet B = \cos\theta\ |A|\, |B|$$

where $|A|$ and $|B|$ are the lengths of A and B, respectively. If the vector B
has unit length, then the dot product gives the projection of A onto B:

$$A \bullet B = \cos\theta\ |A|\,, \text{provided } |B| = 1$$

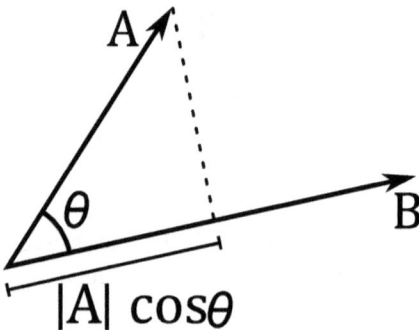

Figure 2. Two vectors A
and B that meet at an
angle θ. The projection
of A onto B is found by
dropping a perpendicular
from the tip of A onto B.

Amazingly, functions behave analogously to vectors if we use integration to define a *dot* or *scalar product* between functions. Let $f(x)$ and $g(x)$ be two continuous, integrable functions. The dot product of $f(x)$ and $g(x)$ is defined as:

$$a = \int f(x)g(x)dx$$

The brilliant, enigmatic, British physicist Paul Dirac[7] invented the following delightful notation for this construction:

$$a = <f(x)|g(x)> = \int f(x)g(x)dx$$

Here $< f(x) \mid$ is called a *bra vector* and $\mid g(x)>$ is called a *ket vector*; together they make a "bracket."

With this definition, angles and magnitudes between functions are analogous to angles and magnitudes between geometric vectors:

$$< f(x) \mid g(x) > = \cos\theta \, \| f(x) \| \, \| g(x) \|$$
$$= \cos\theta \, \| f(x) \|, if \, \| g(x) \| = 1$$

where $\| f(x) \| = <f(x) \mid f(x)>$ and $\| g(x) \| = <g(x) \mid g(x)>$. This allows us to speak in terms of a *function space*, where the component, or projection, of one function onto another is computed using the above expression. An important set of functions in function space are *orthonormal functions*, where all the angles between the functions are 90° and the magnitudes of the functions are set equal to 1.

---

7   Graham Farmelo, in his book, *The Strangest Man: The Hidden Life of Paul Dirac, Mystic of the Atom* (Faber and Faber, 2009), relates the following antidote: At a conference, an attendee raised his hand and said: "I don't understand the equation on the top-right-hand corner of the blackboard." After a long silence, the moderator asked Dirac if he wanted to answer the question, to which Dirac replied: "That was not a question, it was a comment."

Perplexingly, mathematicians and physicists do not use the phrase *function space* to denote the geometry-like aspects of functions. Instead, they call this structure a *vector space*. The uninitiated may think that a "vector space" is a space of vectors, such as the vectors A and B above. But, no, when a mathematician or a physicist says "vector space," he or she is referring to the concept nonexperts would call a "function space."

Group theory enters physics, dominating the search for nature's fundamental laws.

Just as mathematicians use group theory to study nature's geometric symmetries, physicists use group theory to explore symmetries in nature's physical laws. Physics is revealed through group theory. With apologies to William Blake, modern physics is the search for:

> Particle Particle, burning bright,
> In the spectrum of the light;
> What immutable law or rule,
> Could explain thy regular symmetry?[8]

The hydrogen atom is the simplest, most symmetric atom of all, having but a single proton and a single electron. Hydrogen (and other atoms) emit light at discrete frequencies, rather than as a continuous spectrum, as observed in the spectrograph in Figure 3. The hydrogen

---

8   The first stanza of William Blake's poem "The Tyger" is:

> Tyger Tyger, burning bright,
> In the forests of the night;
> What immortal hand or eye,
> Could frame thy fearful symmetry?

spectrum exhibits several bands of light, each with a lengthening arrangement of spaces. A formula for this arrangement was provided in 1889 by the Swedish physicist Johannes Rydberg.[9] It is:

$$\frac{1}{\lambda_{vac}} = R_H \left(\frac{1}{n_1^2} - \frac{1}{n_2^2} \right)$$

where lambda is the wavelength of light emitted, R_H is the Rydberg constant, n_1 is the principal quantum number of an energy level, and n_2 is the principal quantum number of an energy level of the transition state. Here $n_2 > n_1$.

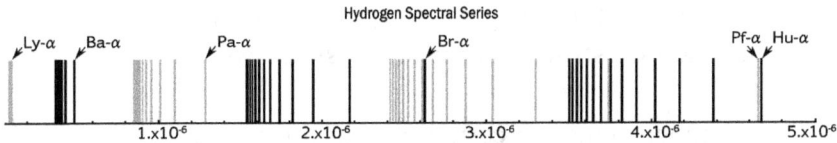

Figure 3. The spectrum of light emitted by the hydrogen atom. Note that this spectrum contains several bands, each beginning at a set frequency, with a lengthening arrangement of spaces.[10]

In 1926, Wolfgang Pauli examined the symmetry of the hydrogen atom.[11] Pauli derived the spectrum of the hydrogen atom by using rotational invariance of the energy, the angular momentum, and the shape of the electron orbits in "function space." Pauli's model of the

9   Johannes Robert Rydberg, "Researches sur la constitution des spectres d'émission des éléments chimiques," *Kongliga Svenska Vetenskapsakademiens Handlingar* 23, no. 11 (1890): 1–177, https://lucris.lub. lu.se/ws/portalfiles/portal/39556483/rydberg_1889_reduced_archived.pdf.
10   Image by Caitlin Jo Ramsay (Wikimedia Commons, CC BY-SA 3.0).
11   W. Pauli, "Über das Wasserstoffspektrum vom Standpunkt der neuen Quantenmechanik," *Zeitschrift für Physik* 36, no. 5 (1926): 336–63, https://doi. org/10.1007/BF01450175.

hydrogen atom gives precisely the spectrum in Figure 3. The Rydberg formula for the hydrogen spectrum is a consequence of the symmetry of the equations.[12] Nature is symmetric, Pauli showed, and the hydrogen spectrum provides exhibit A.

We draw the following conclusions:

1. The hydrogen atom consists of an electron orbiting a proton and obeying the same rotational invariance law as is followed by a planet orbiting the sun.[13]
2. The energy levels of electrons in orbit are quantized. This means that electrons orbit the nucleus with fixed energy levels rather than at any level.
3. Nature is symmetric. Symmetry is a property of some geometrical structures and all physical laws.

These conclusions are certain within the context of the above discussion. Questions remain, though, about where the electrons are in their orbits. Under the Copenhagen interpretation of quantum mechanics, an electron does not exist in particle form, and hence has no specific location, until it is observed.[14] Under the de Broglie–

---

12   Hagen Kleinert, "Group Dynamics of the Hydrogen Atom," in *Boulder Lectures in Theoretical Physics, 1967*, eds. Asim O. Barut et al. (Taylor and Francis 1968), 427–82.

13   The Laplace–Runge–Lenz (LRL) vector describes the shape and orientation of the orbit of one body around another, such as a planet revolving around a star or an electron orbiting a nucleus. The LRL vector is conserved in all problems in which two bodies interact by a central force that varies as the inverse square of the distance between them, such as the gravitational field or the electrostatic field. Pauli used invariance of the LRL vector as a central element in deriving the hydrogen spectrum. See Cornelius Lanczos, *The Variational Principles of Mechanics*, 4th ed. (Dover 1986), 118, 129, 242, 248.

14   Richard Liboff, *Introductory Quantum Mechanics*, 4th ed. (Addison-Wesley, 2002).

Bohm theory, electrons have specific locations, but these are known only statistically. In either case, one can speak of the probability of finding an electron in a specific location around a nucleus, but not of its exact position. The Rutherford–Bohr atomic model is replaced in the modern view by the probability density distribution in Figure 4.

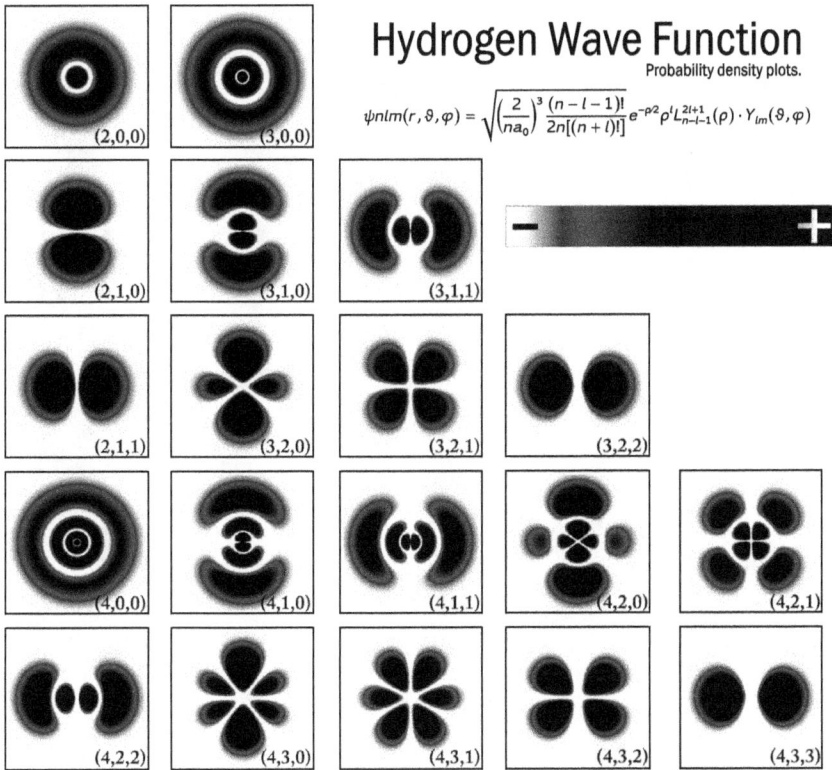

Hydrogen Wave Function
Probability density plots.

$$\psi nlm(r,\vartheta,\varphi) = \sqrt{\left(\frac{2}{na_0}\right)^3 \frac{(n-l-1)!}{2n[(n+l)!]}}\, e^{-\rho^2}\rho^l L^{2l+1}_{n-l-1}(\rho) \cdot Y_{lm}(\vartheta,\varphi)$$

(2,0,0) (3,0,0)
(2,1,0) (3,1,0) (3,1,1)
(2,1,1) (3,2,0) (3,2,1) (3,2,2)
(4,0,0) (4,1,0) (4,1,1) (4,2,0) (4,2,1)
(4,2,2) (4,3,0) (4,3,1) (4,3,2) (4,3,3)

Figure 4. The probability of finding an electron in orbit around a hydrogen nucleus. The three numbers in parentheses (i, j, k) show the energy level of the orbit.[15]

15   Image by PoorLeno.

THE ANSWER

Are the laws of the universe symmetric? Yes, symmetry is a fundamental aspect of the universe. Emmy Noether showed that every conservation law of physics can be derived from the symmetries in the equations.

WHAT ARE THE FUNDAMENTAL COMPONENTS OF MATTER?

NATURE'S BUILDING BLOCKS

Murray Gell-Mann ought to be famous. Not as famous as Newton or Einstein, perhaps, but a recognized name outside of the physics community. Gell-Mann is the originator of the Standard Model of particle physics, a model that has been called "a thing of beauty...incredibly precise and accurate in its predictions."[1] Yet, his fame has not spread. Although Gell-Mann originated the basic concept underlying the Standard Model, his first

---

1   Oscar Miyamoto Gomez, "Five Mysteries the Standard Model Can't Explain," *Symmetry*, October 18, 2018, https://www.symmetrymagazine.org/article/five-mysteries-the-standard-model-cant-explain.

publication, in 1961, provided only a hint of the correct theory.[2] His second publication, in 1964, a theory where Gell-Mann introduced the concept of the quark,[3] was matched by a simultaneous publication by another physicist, George Zweig.[4] Soon, a flood of physicists, Sheldon Glashow, Steven Weinberg, Peter Higgs, Abdus Salam, James Bjorken, and others, filled out the details of the Standard Model. Gell-Mann is lost among all the glory.

> The Standard Model of particle physics
> explains what we call, in layman's terms, the
> fundamental components of matter.

It provides a unified theory of nature's building blocks and their organization. The first thing to know about the Standard Model of particle physics is that physicists do not think of particles the way you or I do, as if an elementary particle was a tiny billiard ball. To a physicist, an elementary particle is a wave, or a field, spread throughout space, except that it interacts with other particles/waves/fields as if it were a particle. A simple example of this is light. Light is a wave traveling through space, but physicists call it a *photon*. Light is a photon because it is emitted and absorbed all at once as a quantum of energy. It is impossible to absorb a half quantum of light.[5]

---

2   Murray Gell-Mann, "The Eightfold Way: A Theory of Strong Interaction Symmetry," Synchrotron Laboratory Report CTSL-20 (March 1961), https://doi.org/10.2172/4008239.

3   Murray Gell-Mann, "A Schematic Model of Baryons and Mesons," *Physics Letters* 8, no. 3 (February 1964): 214–15, https://doi.org/10.1016/S0031-9163(64)92001-3.

4   G. Zweig, "An SU(3) Model for Strong Interaction Symmetry and its Breaking," CERN-TH-401 (January 1964), https://doi.org/10.17181/CERN-TH-401.

5   In the de Broglie–Bohm interpretation of quantum mechanics, the photon is a massless particle that creates an associated electromagnetic field. When the photon is absorbed, the associated field vanishes.

The second thing to know about the Standard Model is that it is based on group theory (we explored group theory in detail in the previous question, *Question 24*). The details of the physics are Hamiltonian, with attendant mathematical complexity, but the top-level view is simplicity itself. Murray Gell-Mann and George Zweig independently defined three "quarks" according to the vectors:

$$\text{up quark} \rightarrow \begin{pmatrix} 1 \\ 0 \\ 0 \end{pmatrix}, \qquad \text{down quark} \rightarrow \begin{pmatrix} 0 \\ 1 \\ 0 \end{pmatrix}, \qquad \text{strange quark} \rightarrow \begin{pmatrix} 0 \\ 0 \\ 1 \end{pmatrix}$$

Of particular importance in physics is the Special Unitary group in three dimensions, labeled SU(3). This group is special because it preserves the energy of the system through rotations, and it is unitary because it preserves magnitudes through rotations. An example of an SU(3) rotation is the matrix:

$$A = \begin{pmatrix} 0 & 1 & 0 \\ -1 & 0 & 0 \\ 0 & 0 & 1 \end{pmatrix}$$

When quarks are multiplied by A, this rotation transforms up quarks into down quarks, and vice versa.

Physicists stopped inventing names and began using ordinary, if whimsical, English words to denote new quantities. The transformation A is called the *flavor* rotation. "Flavor" in the Standard Model has nothing to do with flavor in the ordinary sense of the word.

Under the Gell-Mann and Zweig theories, protons and neutrons are combinations of quarks. A proton consists of two up quarks and one down quark; a neutron consists of one up and two down quarks. See Figure 1. The existence of quarks was validated four years later, in 1968, at the Stanford Linear Accelerator (SLAC), where scattering experiments showed that the proton contained much smaller, point-like objects and was therefore not an elementary particle.[6] Many experiments since that time have confirmed the existence of quarks.[7] Experimental evidence shows quarks are point-like entities, no bigger than 10^{-4} times the size of a proton.

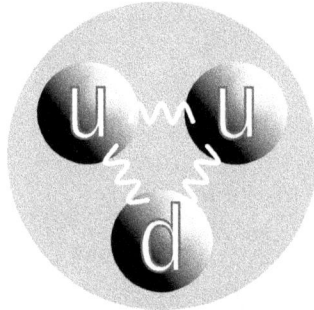

Figure 1. Schematic picture of a proton comprising two up quarks and one down quark. In reality, quarks are thousands of times smaller than protons.[8]

If there were only three quarks and one SU(3) rotation, life would be simple. In this limited universe, the proton and the neutron would be made up of three quarks, electrons would orbit the atomic nucleus, electrons would absorb and emit photons when they change energy levels, and, in this simplistic view, that would be it. However, experiments have revealed two other categories of particles besides the proton and neutron. And theoretical calculations of energy and momentum balance necessitate additional particles as

6   M. Breidenbach et al., "Observed Behavior of Highly Inelastic Electron–Proton Scattering," *Physical Review Letters* 23, no. 16 (October 1969): 935–39, https://doi.org/10.1103/PhysRevLett.23.935.

7   Bill Carithers and Paul Grannis, "Discovery of the Top Quark," *Beam Line* 25, no. 3 (Fall 1995): 4–16, https://www.slac.stanford.edu/pubs/beamline/25/3/25-3-carithers.pdf.

8   Image by Arpad Horvath (Wikimedia Commons, CC BY-SA 2.5).

well. These measurements and theoretical requirements complicate the Standard Model of particle physics.

Before we present the full complexity of the Standard Model, let's consider where we are. Gell-Mann and Zweig's definition of quarks is surprisingly simple. In the Gell-Mann–Zweig model, the most fundamental existents in the universe obey Hamiltonian mechanics, with accompanying complexity, but the interactions between them can be expressed in terms of 1s and 0s. This is a mind-bending discovery. The full Standard Model of particle physics is more complex than the Gell-Mann–Zweig model, but it is based on the same principle and employs the same method. Symmetry rules the universe; the method used to study symmetry is group theory.

Less than a year after the appearance of the Gell-Mann–Zweig model, Sheldon Glashow and James Bjorken predicted the existence of a fourth quark, which they called *charm*. Then, in 1973, Makoto Kobayashi and Toshihide Maskawa proposed the existence of an additional pair of quarks, named the *top* and *bottom* quarks by the Israeli physicist Haim Harari. The result was six "flavors" of quarks: up, down, strange, charm, top, and bottom.[9]

Physicists observed charm quarks in 1974 nearly simultaneously at SLAC and at the Brookhaven National laboratory. In 1977, the team at Fermilab observed the bottom quark. The top quark was not observed until 1995, also at Fermilab. It has an enormous mass, almost as large as that of a gold atom.

Elementary particles include the six quarks, but not the proton and the neutron, that are composed of three quarks each. Other particles that are composed of quarks are lambda, sigma, xi, and omega. In isolation, the only stable particle on this list is the proton. The neutron is stable when bound in a nucleus but has a half-life of

9   Chris Quigg, "Particles and the Standard Model," in *The New Physics for the Twenty-First Century*, ed. Gordon Fraser (Cambridge University Press, 2006), 91.

fifteen minutes in isolation. The other particles on this list have half-lives ranging from 10^{-10} to 10^{-24} seconds.[10]

Electrons are involved in radioactive beta decay. However, the before-and-after states in the decay did not preserve energy and momentum. To make these quantities balance, Wolfgang Pauli proposed in 1930 a new, nearly invisible particle, called the *neutrino*, to carry away the missing energy and momentum. The neutrino is incredibly elusive but was detected in 1956.[11]

Two particles similar to the electron have also been observed. These are the muon and the tauon. Associated with the muon and the tauon are the muon-neutrino and the tauon-neutrino. Together, this group of particles comprise the *lepton* family. All matter is composed of quarks and leptons. Quarks and leptons together are called *fermions*. As indicated in the following Generations of Matter table, three "generations" of fermions exist, with the first generation more stable than the second generation, which is more stable than the third generation.

Generations of Matter				
Fermion categories		**Elementary particle generation**		
Type	*Subtype*	*First*	*Second*	*Third*
Quarks (colored)	**down-type**	down	strange	bottom
	up-type	up	charm	top
Leptons (color-free)	**charged**	electron	muon	tauon
	neutral	electron neutrino	muon neutrino	tau neutrino

10   David Griffiths, *Introduction to Elementary Particles*, 2nd revised ed. (Wiley-VCH, 2008).

11   "The Reines-Cowan Experiments: Detecting the Poltergeist," *Los Alamos Science* 25, no. 3 (November 1997): 4–27, https://www.roma1.infn.it/~maiani/corso-fermi/letture-consigliate/letture-consigliate3/Detecting-the-Poltergeist.pdf.

The Standard Theory contains force-carrying particles. We have met one of these already: the photon. When an electron in an atomic orbit drops from a higher-energy-level orbit into a lower-energy-level orbit, it emits a photon. When an electron in orbit absorbs a photon, it is raised from a lower-energy-level orbit to a higher one. Even though the photon is an electromagnetic wave, the energy associated with the wave is quantized. The photon is a "force mediating particle" with the power of transferring force from one particle to another.[12]

Even before the discovery of quarks, physicists surmised that a strong force must exist to hold the protons in an atomic nucleus together. Like charges repeal and each proton has a positive electric charge. Therefore, a short range but strong force must exist to overcome the electromagnetic repulsive force. To transmit this strong force from one proton to another, physicists hypothesized a massless force-mediating particle called a *gluon*, analogous to the photon that carries the electromagnetic force. After quarks were established, gluons became the particle holding quarks together in protons and neutrons.

Everyone feels gravity, experiences the effects of the electromagnetic force, and understands that a strong force is needed to hold atomic nuclei together. Forces that push and pull things around are eminently logical. But the fourth force of nature, the *weak interaction force*, does not move things. The weak force governs radioactive decay. It converts one flavor of quark into another flavor of quark. For example, in beta decay, the weak force converts a down quark inside a neutron into an up quark, turning the neutron into a proton, simultaneously emitting an electron and an antineutrino. An atomic nucleus with too many neutrons compared to the number of protons transforms itself into a nucleus with one fewer neutron and one

---

12   Gregg Jaeger, "The Elementary Particles of Quantum Fields," *Entropy* 23, no. 11 (2021): 1416, https://doi.org/10.3390/e23111416.

additional proton. Radioactive carbon dating with C^{14} (carbon 14) is possible because of beta decay. The C^{14} nucleus has eight neutrons but only six protons; beta decay turns it into N^{14} (nitrogen 14) with seven neutrons and seven protons. The half-life of this transformation is 5,730 years.[13]

Beta decay, as well as other aspects of the Standard Model, are explained through symmetry. The symmetry, called *electroweak symmetry*, discovered by Sheldon Glashow, Abdus Salam, and Steven Weinberg, who were awarded the 1979 Nobel Prize in Physics for this work. Glashow, Salam, and Weinberg, working separately, each discovered in 1968 that a symmetry group exists that combines electromagnetic interactions and weak interactions into one electroweak interaction. The photon carries the electromagnetic force; three new particles were postulated to carry the weak force. One of the new particles, the Z boson, has zero electric charge; the other two, the W^+ and W^- bosons, carry positive and negative charges, respectively.

The predicted W and Z bosons were discovered at CERN in 1983. These particles are heavyweights, approximately 80 times as massive as the proton. This led to a conundrum: Photons, the particles that carry the electromagnetic force, are massless; gluons, the particles that carry the strong force, are massless; but the W and the Z bosons, the particles that carry the weak force, are massive. How is this possible?

As we saw in *Question 24—Are the laws of the universe symmetric?*, the symmetry of a hexagon under 60° rotations does not depend on the size of the hexagon. We say the 60° rotation operator R is invariant with respect to the gauge we use to measure the hexagon's size. The hexagon may be 1 centimeter, 1 inch, or 1 meter on a side, it does not matter, the operator R stays the same.

---

13   Sylvie Braibant et al., *Particles and Fundamental Interactions: An Introduction to Particle Physics* (Springer, 2011), 313–14.

Gauge invariance is central to physics
just as is symmetry invariance.

Mass was not gauge invariant in the early group-theoretic models of the W and Z bosons.[14] Mass would grow or shrink depending on the scale that was used, an unacceptable result with the first theories proposed. To save gauge invariance, mass must attach to these particles by some other means.

The "other means for acquiring mass" was proposed by three groups of physicists: Robert Brout and François Englert in August 1964;[15] Peter Higgs in October 1964;[16] and Gerald Guralnik, Carl Hagen, and Tom Kibble (GHK) in November 1964.[17] Each group hypothesized a new field and a new particle, separate from the weak interaction field and particles that couple to the W and Z bosons to create mass. Afterwards, Steven Weinberg hypothesized that the new particle was also responsible for providing mass for all the other particles with mass in the Standard Model.[18]

With so many authors, naming the new field and the new mechanism for acquiring mass was unwieldy. Peter Higgs called it the *A-B-E-G-H-H-K-'tH* mechanism [for Anderson, Brout, Englert, Guralnik,

---

14   This is true for quarks and leptons as well.

15   F. Englert and R. Brout, "Broken Symmetry and the Mass of Gauge Vector Mesons," *Physical Review Letters* 13, no. 9 (August 1964): 321–23, https://doi.org/10.1103/PhysRevLett.13.321.

16   Peter W. Higgs, "Broken Symmetries and the Masses of Gauge Bosons," *Physical Review Letters* 13, no. 16 (October 1064): 508–509, https://doi.org/10.1103/PhysRevLett.13.508.

17   G. S. Guralnik et al., "Global Conservation Laws and Massless Particles," *Physical Review Letters* 13, no. 20 (November 1964): 585–87, https://doi.org/10.1103/PhysRevLett.13.585.

18   John Ellis et al., "A Historical Profile of the Higgs Boson," arXiv (January 2012), https://doi.org/10.48550/arXiv.1201.6045.

Hagen, Higgs, Kibble, and 't Hooft],[19] but this name did not catch on. Nowadays, everyone calls it the Higgs field, the Higgs particle, and the Higgs mechanism. The simple one-word name won out over more inclusive variations.[20] Higgs became famous; Brout, Englert, Guralnik, Hagen, and Kibble, not so much.

The Higgs boson thus became the central component of the Standard Model of particle physics, but it had never been seen. The European Organization for Nuclear Research spent approximately $9 billion in an effort to find it.[21] The Large Hadron Collider was constructed at CERN, near Geneva, Switzerland, specifically to either confirm or exclude the existence of the Higgs boson. As pictured in Figure 2, the collider is contained in a circular tunnel, with a circumference of 26.7 kilometers (16.6 miles), at a depth ranging from 50 to 175 meters (164 to 574 feet) underground. The 3.8-meter (12-foot) wide tunnel was originally constructed to house an electron-positron collider. Protons in the newer hadron collider move at about 0.999999990 c, or about 3.1 m/s (11 km/h) slower than the speed of light. Two beams of protons are created, traveling in opposite directions. These beams collide at four locations around the ring. Protons are bunched together, into 2,808 bunches, with 115 billion protons in each bunch, to produce a high collision rate at the detectors.

On July 4, 2012, CERN announced they had discovered a previously unknown boson with the expected mass. The probability of detecting particles of these characteristics by chance alone is less than one in three million. On March 14, 2013, CERN issued the following statement:

---

19   Frank Close, *The Infinity Puzzle: Quantum Field Theory and the Hunt for an Orderly Universe* (Oxford University Press, 2011).
20   Peter Watkins, *Story of the W and Z* (Cambridge University Press, 1986), 70.
21   "LCH: un (très) petit Big Bang," Agence Science-Press, November 25, 2009, https://www.sciencepresse.qc.ca/actualite/2009/11/25/lhc-petit-big-bang.

Figure 2. The Large Hadron Collider at CERN sits in a circular tunnel 26.7 kilometers in circumference and 100 meters underground.[22]

[Two major experiments at CERN] have compared a number of options for the spin-parity of this particle, and these all prefer no spin and even parity [two fundamental criteria of a Higgs boson consistent with the Standard Model]. This, coupled with the measured interactions of the new particle with other particles, strongly indicates that it is a Higgs boson.[23]

With confirmation that the Higgs particle exists, the Nobel Committee awarded Englert and Higgs the Nobel Prize in Physics

---

22   Photo by Samuel Joseph Hertzog, CERN. Reproduced with permission from CERN.
23   CERN, "New Results Indicate that New Particle Is a Higgs Boson," news release, March 14, 2013, https://home.web.cern.ch/news/news/physics/new-results-indicate-new-particle-higgs-boson.

in 2013. Why the Nobel Committee snubbed Guralnik, Hagen, and Kibble is controversial.[24]

A schematic of the complete Standard Model of particle physics is presented in Figure 3. The outermost ring in this schematic provides four groups of three mass-carrying particles. The four lowest mass particles in each group are given first: the electron, the up quark, the down quark, and the electron neutrino. Each of these particles has two heavier siblings. Inside the mass-carrying ring are the four force-carrying particles: the photon (electromagnetic force), the gluon (the strong nuclear force), and the W and Z bosons (the weak nuclear force). Finally, in the center, there is the Higgs boson, the particle that gives mass to the universe.

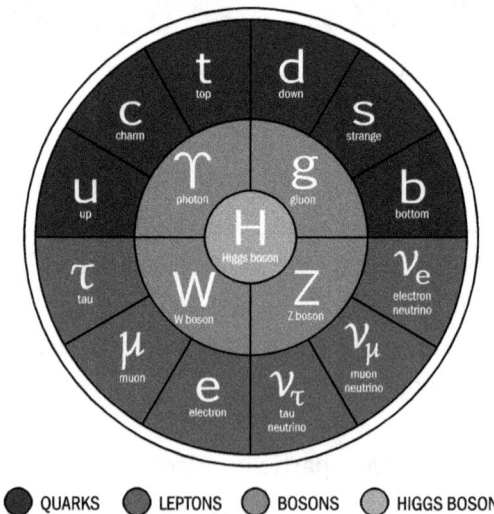

Figure 3. Schematic diagram of the seventeen elementary particles in the Standard Model of the universe. The outer ring contains the twelve particles that carry mass, the inner ring four particles that carry force, and the Higgs Boson, the particle that provides the mass for the twelve outer-ring particles, is in the center.[25]

● QUARKS ● LEPTONS ● BOSONS ○ HIGGS BOSON

---

24   Brout had died in 2011. T. C., "Why Are Some Scientists Unhappy with the Nobel Prizes," *The Economist*, October 10, 2013, https://www.economist.com/the-economist-explains/2013/10/10/why-are-some-scientists-unhappy-with-the-nobel-prizes.

25   Image courtesy of *Symmetry Magazine*, a joint Fermilab/SLAC publication. Artwork by Sandbox Studio, Chicago.

The Standard Model's predictions of the existence and properties of elementary particles have passed test after test. The first test was the discovery of quarks with the properties predicted by theory. Next came the prediction and subsequent observation of the W and Z bosons. While the mass of most elementary particles can only be measured and accepted, not so with the W and Z bosons. The electroweak theory predicts the mass of the W boson through other measurable quantum properties. The predicted value prior to 2022 was 80.390 ± 0.018 GeV; the measured value is 80.387 ± 0.019 GeV. For the Z boson, the theory predicts a mass of 91.1874 ± 0.0021; the measured value is 91.1876 ± 0.0021 GeV. The prediction and discovery of the Higgs boson further confirmed the correctness of the Standard Model.

The spectacular success of the Standard Model over the past five decades has overshadowed its weaknesses. The most glaring weakness is the absence of gravity in the Standard Model. Physicists have hypothesized a particle called the *graviton*, which carries the gravitational force, but they have never observed it. Other weaknesses include the lack of particles to account for dark matter and dark energy, two forces surmised from galactic-scale behavior. Although alternative explanations of dark matter and dark energy are possible, the currently accepted view is that the Standard Model is incomplete without them.[26]

Then, in 2022, something went amiss. Prior to building the Large Hadron Collider at CERN, the Tevatron, a 6-kilometer underground ring at Fermilab, near Chicago, Illinois, was the largest particle accelerator in the world. The Tevatron shut down in 2011 after the completion of the much more powerful Large Hadron Collider. But, from 1983 to 2011, during its years of operation, the Tevatron had recorded data from approximately four million W boson candidates. With so

---

26   Robert Oerter, *The Theory of Almost Everything: The Standard Model, the Unsung Triumph of Modern Physics* (Penguin, 2006), 2.

much recorded data, and no new data coming in, the 398 scientists associated with Fermilab went through this old data to get a more precise result for the W boson mass. The W boson mass measured by the Fermilab scientists is 80,433.5 ± 9.4 million electron volts (MeV) or 0.1 percent higher than the value predicted by the Standard Model.[27]

A 0.1 percent discrepancy between measurement and simulation does not seem like much, but the Fermilab publication has shaken the physics community to its foundations. The 0.1 percent discrepancy means that either the measurement or the theory is wrong. And the Fermilab measurement was performed with extreme care.

DO QUARKS MATTER?

While the Standard Model provides our best understanding of the fundamental existents in the universe, it has not made an impression on the public mind. Electrons, protons, and neutrons enter our daily lives through chemistry and technology. Most people are aware of the atomic theory of matter and of the electron's role powering our lives. But few people are mindful of muons, quarks, and bosons. Indeed, some authors, such as Kary Mullis, Nobel laureate in Chemistry, believe that elementary particles are not important to us. Mullis writes:

> Do we need to use billions of dollars to build machines that maybe will put a few of our rightfully treasured eggheads in touch with things so far from what can be engineered into useful items that only they will get a thrill out of finding them?... When did we as a culture decide that extremely little things were fundamental? I

27   CDF Collaboration, "High-Precision Measurement of the W Boson Mass with the CDF II Detector," *Science* 376, no. 6589 (April 2022): 170–76, https://doi. org/10.1126/science.abk1781.

think it was this century and the advent of nuclear bombs. At the same time, we decided that very big things were also important. Medium sized things like us were relegated to the not-so-important closet.[28]

We should not spend billions of dollars on fundamental research, if the "we" implies the initiation of force by the government. However, this is not Mullis's point. Mullis is decrying the study of "little things" when he received the Nobel Prize for developing the polymerase chain reaction (PCR) used in DNA analysis.[29] The irony is evident: DNA is tiny compared to "medium sized things like us"; Mullis appears to be more concerned with "what can be engineered" than with size per se.[30]

Of the exotic particles in the Standard Model, particles that are unstable and/or short lived, only the positron has found commercial application. Positron emission tomography (PET) employs positrons to scan patients for cancer. In a PET scan, a glucose solution containing a radioactive isotope is injected into a cancer patient. The radioactive isotope is obtained by using a particle accelerator like the one at CERN, but much smaller. The injected glucose and associated radioisotope become concentrated in the fast-growing cancer cells. The radioisotope emits a positron, the antiparticle of the electron. The emitted positron travels for a short distance, during which it decelerates to a point where it can interact with an electron. The encounter annihilates both the electron and the positron, producing

28   Kary Mullis, *Dancing Naked in the Mind Field* (Pantheon, 1998), 70–71.

29   PCR employs the ability of a section of DNA to copy itself over and over, first two copies, then four, then eight, and so on, until a large quantity of the DNA section is produced.

30   Emily Yoffe, "Is Kary Mullis God? (Or Just the Big Kahuna?)," *Esquire* 122, no. 1 (July 1994): 68–75, https://classic.esquire.com/article/1994/7/1/is-kary-mullis-god-or-just-the-big-kahuna.

a pair of photons moving in opposite directions. These create two bursts of light on opposite sides of the patient, detected by a scanning device. Photons that do not arrive in temporal pairs are ignored. Computer software is then used to convert the millions of flashes detected by the scanner into a three-dimensional picture of the cancer inside the patient.[31]

Whether muons, quarks, and bosons will ever find commercial applications is impossible to predict. When Thomson discovered the electron in 1897, he could not have imagined that his electron beam would make television sets forty years later. Similarly, when Rutherford discovered the proton and hypothesized the neutron in 1917, he did not know an atomic bomb would be one result of his work.

Perhaps, one day, we will harness the quark.

THE ANSWER

What are the fundamental components of matter? Quarks, leptons, and bosons make up matter. These particles were first theorized through symmetry—the idea that nature's laws are symmetric. Now called the Standard Model of particle physics, experiments have confirmed the existence of these particles to a high degree of accuracy.

---

31 Michael E. Phelps, ed., *PET: Physics, Instrumentation, and Scanners* (Springer, 2006), 8–10.

WHAT IS LIGHT?

THE GENIUS OF LIGHT: JAMES CLERK MAXWELL

Electricity mystified men from ancient times:

> The ancient Tuscans by their learning hold that there are nine gods that send forth lightning and those of eleven sorts.[1]

Written by Pliny the Elder circa 77 CE, these words capture ignorant man's superstition and dread of electricity. Lightning, St. Elmo's fire, and electric eels, all shocking things, were the province of the gods.

In 1785, Charles-Augustin de Coulomb set out to demystify electricity. Coulomb used a torsion balance to quantify how the force between two electrified bodies varies with respect to the distance

---

1   William Maver Jr., "Electricity, its History and Progress," *The Encyclopedia Americana: A Library of Universal Knowledge*, vol. X (Encyclopedia Americana Corporation, 1918), 171.

between them. This balance was based on the principle that the torsional force on a metal wire is proportional to the torsion angle. Coulomb showed that "the repulsive force that the two balls—[which were] electrified with the same kind of electricity—exert on each other, follows the inverse proportion of the square of the distance."[2]

> As with Newton's Law of Universal Gravitation, the inverse square relationship of the electric force with distance is a consequence of the three-dimensional nature of space.

The electric field produced by an electrified body spreads out uniformly over a larger and larger sphere as you move away from the body. Since the surface area of a sphere is proportional to the square of the radius of the sphere, the intensity of the electric field drops off inversely with the square of the distance from the source. This property of the electric field is now called Coulomb's law.

Fear evaporates with understanding, and the Enlightenment produced a lot of understanding. Next in line was another Frenchman, André-Marie Ampère, who measured the force between two long, straight, parallel current-carrying wires in the 1820s. He had been inspired by a discovery by the Danish physicist Hans Christian Ørsted, who noticed that a magnetic needle was deflected by an adjacent electric current. Ampère determined the force per unit length of wire was proportional to the product of the currents flowing in the wires divided by the distance between them.[3] This relationship is called Ampère's law.

---

2   C. A. Coulomb, "Premier mémoire sur l'électricité et le magnétisme," *Histoire de l'Académie Royale des Sciences* (1785): 569–77.

3   André Koch Torres Assis and J. P. M. C. Chaib, *Ampère's Electrodynamics: Analysis of the Meaning and Evolution of Ampère's Force Between Current Elements, Together with a Complete Translation of his Masterpiece: Theory of Electrodynamic Phenomena, Uniquely Deduced from Experience* (Apeiron, 2015).

A straight current-carrying wire provides a cylindrical geometry. Since the surface area of a cylinder is directly proportional to the cylinder radius, the magnetic field decreases in direct proportion to the distance from the wire. Moving a magnetized needle in a circular loop around the wire shows that the magnetic field is always pointing in a direction tangent to the circle. Regardless of the size of the circle you take, the sum (integral) of the magnetic field around a loop is always proportional to the current enclosed by the loop.

Ørsted and Ampère working independently showed that a current-carrying wire produced a magnetic field. In 1831, Michael Faraday discovered the physical effect in the other direction, that a magnetic field induces a current in a wire. In this case, the wire must experience a time-varying magnetic field. The simplest way to do this is to spin a loop of wire in a constant magnetic field. Faraday's discovery led to the development of electric power generators.[4] Almost all the electric power we use today is generated in this way. Faraday's law of induction is like Ampère's law: Regardless of the size of the circle you take, the sum (integral) of the electric field around a loop is always proportional to the time rate of change of the magnetic field enclosed by the loop.

Faraday is an unlikely scientific hero. Born in 1791 to an impoverished family, he had no formal schooling. At fourteen, he became an apprentice in a bookbinder's shop, where he began to read books. At age twenty, he attended lectures by the eminent English chemist Humphry Davy and kept excellent notes. He converted these notes into a three-hundred-page book he presented to Davy. Davy was so impressed by Faraday's work that he hired Faraday to be his assistant

4   "Michael Faraday's generator," The Royal Institution, accessed December 11, 2024, https://www.rigb.org/explore-science/explore/collection/michael-faradays-generator.

at the Royal Institution. Faraday remained at the Royal Institution for the rest of his productive life.[5] He turned down offers to become the President of the Royal Society twice.[6]

Faraday was an experimental physicist with no mathematical training. He understood electromagnetics only through the hundreds of experiments he had performed. This set the stage for a Scottish physicist, James Clerk Maxwell, one of the best mathematical physicists of all time, to express Faraday's and others' experimental results in mathematical form. Today, Maxwell, Newton, and Einstein are generally regarded as the three greatest physicists who shaped our understanding of time, space, and light.

Coulomb's law states that the electric field from a charged body decays as the inverse square of the distance from the charge. Magnetic charges do not exist, an observation we will explain when answering *Question 27—Are space and time linked?*, but a similar law, Coulomb's law for magnetic flux, maintains the same rate of decay for magnetic flux as is the case for electric flux. As we have already explained, the remaining two laws of electric and magnetic fields assume similar forms. The most significant difference is that Faraday's law states that a time-varying magnetic field creates an electric field, while Ampère's law has no corresponding term.

In 1864, Maxwell asked himself: What if I add a term to Ampère's law similar to the one in Faraday's law? The answer he found was revolutionary. Adding this new term to the magnetic field equations, a term now called the *displacement current*, coupled the two sets of equations together. This coupling produced a wave, a wave in which energy oscillates between the electric and magnetic fields.

---

5   James Hamilton, *A Life of Discovery: Michael Faraday, Giant of the Scientific Revolution* (Random House, 2004).

6   John Meurig Thomas, *Michael Faraday and The Royal Institution: The Genius of Man and Place* (Taylor & Francis, 1991).

Evaluating the known constants in the equation, Maxwell determined that the velocity of the waves was equal to the velocity of light.

Maxwell hypothesized that light is an electromagnetic wave.

A generation later, in 1889, Heinrich Hertz demonstrated the existence of Maxwell's hypothesized electromagnetic waves. He showed a spark produced by a discharging capacitor produced a spark in an unconnected capacitor across the room. Hertz measured the properties of the propagating waves and concluded that the waves obeyed Maxwell's equations.

Six years later, in 1895, Guglielmo Marconi used Hertz's spark-gap radiation system to transmit telegraph signals through space. Marconi is generally credited with inventing the radio, but Marconi's spark-gap transmission system was limited to dots and dashes. Radio as we know it, with the ability to carry the human voice, was invented by Nikola Tesla. Tesla created electronic circuits that transmit signals at a specific frequency, and receivers tuned to the same frequency to detect them.

Although Tesla invented the oscillating circuits that made radio possible, he did not realize the business opportunity for his invention was in communications. Tesla had worked in power generation and distribution his entire life and wanted to use radio waves to transmit and distribute power without wires. He spent a fortune developing a wireless power transmission system, to no avail. Tesla, the inventor of the alternating current power system we use today, and the inventor of the oscillating circuits at the heart of radio, died penniless in a New York City hotel room in 1943.

Computer simulations allow us to visualize the "mysterious force" called electromagnetism. Step one in this process is to divide the object and the space around the object to be analyzed into thousands or even millions of small pieces called *finite elements*. An example of such a finite element mesh is presented in Figure 1A. This figure shows the tessellation of a four-propeller drone using tetrahedral finite elements. Only one layer of the finite element mesh is displayed to make the elements in this layer visible. Once the mesh is formed, an approximation to Maxwell's equations is made in each element and the approximations combined into a large matrix equation. This equation is then solved for the approximate values of the electromagnetic field in each element. The resulting approximate electromagnetic field radiating from the drone's antenna can then be displayed on a computer screen, as shown in Figure 1B.

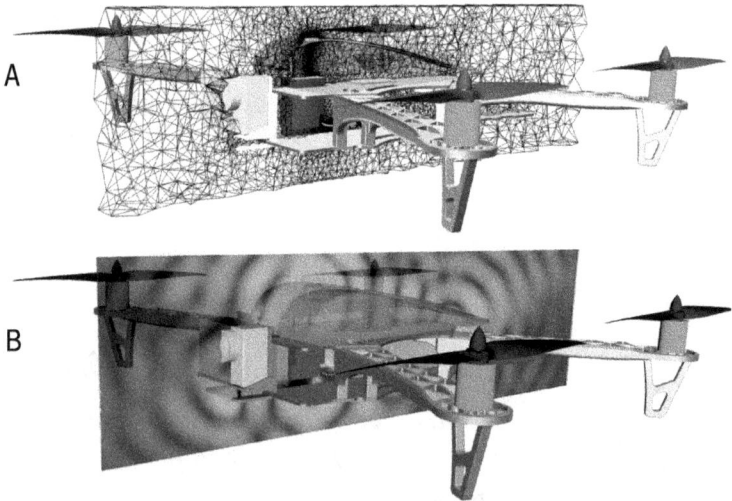

Figure 1. Computer simulation of the electromagnetic radiation from an antenna mounted on a four-propeller drone. (A) The finite element mesh used in the simulation. Everything is broken into a tetrahedral mesh: the electronics, the drone components, and the surrounding space. To make visualization easier, only one layer of the mesh is displayed. (B) The electromagnetic radiation emanating from the drone at one instant of time. Darker regions indicate a higher magnitude of radiation and lighter regions indicate a lower magnitude of radiation. The wave moves outward as time progresses.[7]

7   Figures courtesy of Ansys Inc.

THE ANSWER

What is light? Light is time-varying electric and magnetic fields coupled together with the decreasing energy in the oscillating electric field transferred to an increasing magnetic field, and vice versa.

ARE SPACE AND TIME LINKED?

SPACE-TIME

Space is axiomatic. Time is axiomatic. In our ordinary lives, space and time appear to be two separate things. It took a leap of amazing intellectual ability to show that space and time are coupled, not independent, axioms.

The story begins with the Scottish physicist James Clerk Maxwell, who proposed that light is an electromagnetic wave, as described in answering *Question 26—What is light?* The equations Maxwell derived—now known as Maxwell's equations—describe macroscopic electromagnetic phenomena with remarkable precision. These equations are based on experimental observations by the French scientists Charles-Augustin de Coulomb and André-Marie Ampère and the English scientist Michael Faraday.

In 1905, a young Albert Einstein noticed that Maxwell's equations are not invariant with respect to velocity.

Two people passing each other at high speeds would not see equivalent electric and magnetic fields. Others had noticed this enigma, but Einstein was the first to come up with a solution. He proposed that the velocity of light is invariant for all observers in the universe—i.e., the speed of light is fixed at a constant value regardless of how fast an observer is moving. This solves the enigma produced by Maxwell's equations but leads to unexpected results. When the velocity of light is fixed, time and space are coupled, as is mass and energy. Our conception of time, space, matter, and energy is forever changed.

Newton's view of space and time, and Einstein's view of space and time differ in one essential respect. In Newton's view, the relative motion of two observers makes no difference to space and time; in Einstein's view, space and time appear to be different to the two observers if their relative velocity is a significant fraction of the velocity of light.

All motion in Newton's absolute space is relative with no central or preferred point. Newton assumed that three-dimensional space is filled with an infinite number of fixed (in time) points and that time passes with equal swiftness at all points in this infinite universe. Uniform motion is defined as a linear relationship between space and time. Assume that we are sitting at the origin of a coordinate system we have set up in this infinite space and that a second observer, let's call him Ben, passes by us at time $t = 0$ in the x-direction with velocity v. Time passes as usual for us, as we remain seated at the origin of the coordinate system. The four parameters giving our location in time and space are $(t = t, x = 0, y = 0, z = 0)$.

Ben's location with respect to the x coordinate in our coordinate system is given by:

$$x = vt$$

Ben's y and z coordinates are unchanged:

$$y = 0$$
$$z = 0$$

Ben places himself at the origin in his own world. Let's call the four parameters giving time and space in Ben's world t', x', y', and z'. Since Ben is stationary with respect to his own coordinate system, the four parameters giving Ben's location are ($t' = t'$, $x' = 0$, $y' = 0$, $z' = 0$).

In Newtonian space, three of the four parameters, $t' = t$, $y' = y$, $z' = z$, are the same for both us and for Ben. The parameter x', however, is different. Ben is looking at us retreating in the negative x direction as time passes:

$$x' = -vt'$$

Collecting these four variables together in a single expression provides what is called the Galilean transformation:

$$\begin{bmatrix} t' \\ x' \\ y' \\ z' \end{bmatrix} = \begin{bmatrix} 1 & 0 & 0 & 0 \\ -v & 0 & 0 & 0 \\ 0 & 0 & 1 & 0 \\ 0 & 0 & 0 & 1 \end{bmatrix} \begin{bmatrix} t \\ x \\ y \\ z \end{bmatrix}$$

Each row in this matrix equation is evaluated by multiplying like terms. For example, the top row yields:

$$t' = \begin{bmatrix} 1 & 0 & 0 & 0 \end{bmatrix} \begin{bmatrix} t \\ x \\ y \\ z \end{bmatrix} = 1 \bullet t + 0 \bullet x + 0 \bullet y + 0 \bullet z = t$$

Similarly with the other rows.

The Galilean transformation worked for two hundred years. Then came Maxwell's equations. The Galilean transformation does not make sense when applied to electromagnetics.[1] Suppose you are at rest with respect to the coordinate system as before, but this time you place a charge Q at the origin. You measure the electric field from this charge and discover it is an electric field governed by Coulomb's law. Ben, however, is traveling with velocity v when he measures the electromagnetic field. Ben observes both an electric and magnetic field around a moving charge, with the magnetic field arising from the current generated by the charge's movement, in line with Ampère's law.[2] Ben's relative movement with respect to you has created an entirely new field, the magnetic field you do not see at all. How can this be?

Physicists call a physical law *invariant* under a transformation if the physical law is the same both with and without the transformation. Newtonian mechanics is invariant under the Galilean transformation because Newton's laws of motion are the same for a stationary and a moving observer.[3] The conundrum faced by physicists at the end of the nineteenth century was that Maxwell's equations are not invariant with respect to the Galilean transformation.

Maxwell's equations model electromagnetic
phenomena with great accuracy, but
these equations are incompatible with the
Galilean coordinate transformation.

---

1   David J. Griffiths, *Introduction to Electrodynamics*, 3rd ed. (Pearson Education, 2003).
2   The total force produced by a moving charge is called the Lorentz force and is the superposition of electric and magnetic forces.
3   By "stationary" we mean with respect to an arbitrary coordinate system. All motion is relative. There is no absolute "stationary point" in the universe.

In 1892, the Dutch physicist Hendrik Lorentz discovered a coordinate transformation that made Maxwell's equations invariant with respect to motion.[4] Called the Lorentz transformation, it is given as:

$$\begin{bmatrix} t' \\ x' \\ y' \\ z' \end{bmatrix} = \begin{bmatrix} \gamma & -\frac{\gamma}{c^2}v & 0 & 0 \\ -\gamma v & \gamma & 0 & 0 \\ 0 & 0 & 1 & 0 \\ 0 & 0 & 0 & 1 \end{bmatrix} \begin{bmatrix} t \\ x \\ y \\ z \end{bmatrix}$$

where c is the free space velocity of light and γ is the Lorentz factor:

$$\gamma = \left(\sqrt{1 - \frac{v^2}{c^2}} \right)^{-1}$$

Notice that $\gamma=1$ if the velocity v is small compared to the velocity of light. However, if the relative velocity v is a significant fraction of the velocity of light, γ has a nontrivial value less than 1.

Evaluating the top row as before gives:

$$t' = \begin{bmatrix} \gamma & -\frac{\gamma}{c^2}v & 0 & 0 \end{bmatrix} \begin{bmatrix} t \\ x \\ y \\ z \end{bmatrix} = \gamma \bullet t - \frac{\gamma}{c^2}v \bullet x + 0 \bullet y + 0 \bullet z = \gamma \left(t - \frac{v}{c^2} \right)$$

By this result, t' is always less than t. Therefore, *time slows down* as the relative velocity v approaches the speed of light. Similarly with the second row:

$$x' = \begin{bmatrix} \gamma v & \gamma & 0 & 0 \end{bmatrix} \begin{bmatrix} t \\ x \\ y \\ z \end{bmatrix} = \gamma v \bullet t + \gamma \bullet x + 0 \bullet y + 0 \bullet z = \gamma(vt + x)$$

Moving bodies contract in the direction of motion ($x' < x$) as the velocity of an object approaches the speed of light. A further

4   Hendrik Antoon Lorentz, *La théorie electromagnétique de Maxwell et son application aux corps mouvants* (E. J. Brill, 1892).

implication of this transformation is that the inertial mass of a moving object increases as its velocity approaches the speed of light.[5]

Although Lorentz discovered the transformation at the heart of special relativity, he did not accept it. Philosophically, Lorentz was wedded to Newton's absolute time and space. He always spoke of "local time" and "local space" when discussing time and space for a moving object. It remained for Einstein to dismiss the notion of absolute time and absolute space. In Einstein's universe, time does not flow at the same rate for all observers, and space does not have the same dimensions for all observers. In Newton's universe, simultaneity of events is easily determined by synchronizing clocks. In Einstein's universe, the simultaneity, as a real physical relation among events, is nonexistent. As Maudlin notes:

> Once we abandon Newtonian absolute time and the persistence of points of Newtonian absolute space, there are no objective speeds, either of light or of anything else.[6]

Einstein published his famous paper on special relativity in 1905. Five years later, Lorentz wrote:

> If one connects with this the idea...that space and time are completely different things, and that there is a "true time"..., then it can be easily seen that...true time should be indicated by clocks at

---

5   Hendrik Antoon Lorentz, "Electromagnetic Phenomena in a System Moving with Any Velocity Smaller Than That of Light," *Proceedings of the Royal Netherlands Academy of Arts and Sciences* 6 (1904): 809–831, https://dwc.knaw.nl/DL/publications/PU00014148.pdf. In 1953, Einstein said of Lorentz: "For me personally [Lorentz] meant more than all the others I have met on my life's journey." Quoted in Justin Wintle, ed., *Makers of Nineteenth Century Culture: 1800–1914* (Routledge, 1982), 375.

6   Maudlin, *Philosophy of Physics*, 121.

rest in the aether. However, if the relativity principle had general validity in nature, one wouldn't be in the position to determine, whether the reference system just used is the preferred one. Then one comes to the same results, as if one...denies the existence of the aether and of true time, and to see all reference systems as equally valid. Which of these two ways of thinking one is following, can surely be left to the individual.[7]

All ideas, by necessity, are "left to the individual." The choice we are presented with is:

1. Stick with our intuitive notions of absolute space and time and ignore the contradictions arising from Maxwell's equations.
2. Accept the validity of Maxwell's equations and redefine space and time to be consistent with these equations.

Lorentz persevered with the primacy of consciousness point of view. "Relativity can't be correct," Lorentz appeared to say. "My intuition about space and time doesn't allow it."

> Einstein rejected intuition and placed the primacy
> of existence at the center of his theory.

His reasoning appeared to be, "My job as a scientist is to take existence, measure it, and derive ideas based on these measurements. The measured data from electromagnetics requires that

---

7   Hendrik Antoon Lorentz, "Das Relativitätsprinzip und seine Anwendung auf einige besondere physikalische Erscheinungen," in *Das Relativitätsprinzip: Eine Sammlung von Abhandlungen* ed. Otto Blumenthal (B. G. Teubner, 1913), 74–89.

space and time be modified according to the Lorentz transformation. The results are non-intuitive but consistent with observation and measurement."

The payoff for abandoning absolute space and time goes far beyond explaining the behavior of electric and magnetic fields. The Lorentz transformation implies that mass and energy are related through the famous equation:

$$E = m c^2$$

The conversion of mass into energy, and vice versa, is not found in a universe governed by the Galilean transformation of space and time. Mass-energy equivalence is a central concept in modern physics. Experiments have confirmed this equivalence thousands of times. The conversion of mass into energy explains how the sun has shone with such brilliance for 4.8 billion years.[8] This conversion explains how one element can be transmuted into another. And it explains how the 64 kilograms of uranium 235 in the Little Boy atomic bomb exploded over Hiroshima on August 6, 1945, with the equivalent power of 15,000 tons of TNT.[9]

The mystery of a stationary charge producing only an electric field while a charge moving relative to an observer produces *both* an electric and a magnetic field is solved. Space and time for fast-moving objects appear to be distorted as seen by a stationary observer. The outward-pointing electric field created by a stationary positive charge appears to be a circling magnetic field for a moving charge.

---

8   The Sun's core fuses about 600 million tons of hydrogen into helium every second, converting 4 million tons of matter into energy every second as a result. Carlo Broggini, "Nuclear Processes at Solar Energy," arXic (August 2003), https://doi.org/10.48550/arXiv.astro-ph/0308537.

9   Richard Rhodes, *The Making of the Atomic Bomb* (Simon & Schuster, 1987).

This field is exactly the one computed mathematically using special relativity for moving charges. When you throw a switch to turn a light bulb on, the switch closes the circuit and allows current to flow. The velocity of this current pulse may be computed by using a set of equations called the *transmission line equations*, derived from Maxwell's equations under the condition that the wire is straight and long. The velocity of a pulse traveling down the transmission line depends on the geometry and materials forming the transmission line but is typically 90 percent to 99 percent of the free-space velocity of light. The electric and magnetic fields, as well as the charges and current in the wire, act as a single system. *The free electrons in the wire must support this relativistic electromagnetic wave.* While individual electrons are repeatedly slowed down by collisions with atoms, the free electrons in the wire accelerate to relativistic velocities to provide the relativistic pulse of current traveling down the transmission line. No magnetic charges (sometimes called magnetic monopoles) exist. The magnetic field surrounding a current-carrying wire is the result of space and time being warped by the relativistic velocities of the electrons carrying the current.[10]

THE ANSWER

Are space and time linked? Yes, space and time are linked. The observed behavior of the electromagnetic field as described by Maxwell's equations requires that they be linked. This linkage implies that mass can be converted into energy, and vice versa.

---

10   Magnetic fields from permanent magnets are due to the alignment of the electron's intrinsic magnetic dipole moments. This dipole moment originates from the electron's quantum mechanical spin. Experimental evidence for magnetic monopoles (analogous to electric charges for electric fields) does not exist. Arttu Rajantie, "The Search for Magnetic Monopoles," *Physics Today* 69, no. 10 (2016): 40, https://doi.org/10.1063/PT.3.3328.

IS GRAVITY AN ILLUSION?

GOODBYE, GRAVITY

Tobias Bothwell and a group at the National Institute of Standards and Technology (NIST) in Boulder, Colorado, have developed a clock with mind-blowing accuracy. Using arrays of intersecting laser beams, their optical atomic clock holds roughly 100,000 atoms of ultracold strontium 87 in a lattice, much like eggs are cradled in an egg carton. The strontium atoms oscillate between two states with incredible precision compared to the grandfather clocks of old. NIST's optical clock has an accuracy of twenty-one significant figures. It would lose only a fraction of a second over the entire lifetime of the universe.[1]

---

1   Tobias Bothwell et al., "Resolving the Gravitational Redshift across a Millimetre-Scale Atomic Sample," *Nature* 602 (February 2022), 420–24, https://doi.org/10.1038/s41586-021-04349-7.

> It is not obvious that time has anything to do with gravity, but that's what Einstein's general theory of relativity says.[2]

According to the general theory, gravity slows time down, just like a runner is slowed by wading deeper into a river. The stronger the gravity, the slower the clock ticks. With the NIST clock, lowering it a mere 0.2 millimeters toward the center of the earth, roughly twice the thickness of a sheet of paper, slows the clock down by a measurable amount. The measured time dilation agrees with the prediction of the general theory of relativity.

Einstein developed his second mind-expanding worldview by revisiting Galileo's most famous experiment. Galileo is said to have dropped two cannonballs of different weights from the Leaning Tower of Pisa. Aristotle had argued that the heavier ball falls faster than the lighter ball. Galileo showed that the two balls fall at the same speed.[3]

Einstein said, in effect, "Put an accelerometer in one cannonball and see what happens." The accelerometer may be realized as a vertical structure containing a mass m with two springs attached, one above, the other below the mass. At the top of the Tower, while Galileo is holding the ball, the upper spring is stretched, the lower spring compressed, by the force of gravity. Thus, while Galileo is holding the ball, the accelerometer is showing an acceleration. After Galileo drops the cannonball, the springs return to their neutral positions and show zero acceleration. While the cannonball appears to be accelerating in Galileo's frame of reference, the accelerometer inside the cannonball says there is no acceleration.

2   Albert Einstein, "Die Grundlage der allgemeinen Relativitätstheorie," *Annalen der Physik* 354, no. 7 (1916): 769–822, https://doi.org/10.1002/andp.19163540702.
3   Most historians doubt that Galileo actually performed this experiment physically. Robert P. Crease, "The Legend of the Leaning Tower," *Physics World*, February 4, 2003, https://physicsworld.com/a/the-legend-of-the-leaning-tower/.

Einstein argued that the falling cannonball, not
Galileo, is the one at rest in space-time.

Space-time is curved by gravity and the cannonball is moving
along a straight-line path in space-time without any force being
applied to it. It is Galileo, according to Einstein, who is accelerating
by pushing with his feet against the top of the Tower.

A free-falling accelerometer experiences no force. This led
Einstein to formulate the Strong Equivalence Principle. According
to this principle, an experiment performed "in free fall" in a uniform
gravitational field will have the same result as if performed in an
inertial laboratory in empty space. Conversely, an experiment performed in a laboratory experiencing a uniform gravitational field
will produce the same result as performed in a uniformly accelerating laboratory in empty space.

We have all seen astronauts in the International Space Station
free-floating in space as they circle the earth. This is partial experimental verification of the Strong Equivalence Principle. Einstein's
genius was to take this relatively simple concept and derive the
equations of space and time from it. A consequence of these equations is that space-time is influenced by the distribution of matter
within it, but space-time is not determined by it.

Special relativity is based on the postulate that the laws of electromagnetics, as described by Maxwell's equations, are the same
regardless of the relative velocity of the observer. While this postulate provides results that are nonintuitive, it has been confirmed
experimentally thousands of times. General relativity is based on a
very different postulate, that inertial mass and gravitational mass
are the same. This postulate is intuitively more appealing than the
special relativity postulate, but has even more dramatic implications.
The very nature of space and time is described by the general theory
of relativity. According to general relativity, space is both finite and

unbounded, analogous to the surface of a sphere, which has a finite area but no edges.

The beauty of the general theory is that gravity disappears. The force of gravity is replaced by the curvature of space-time. With respect to objects containing mass, general relativity provides a structure for space and time just as special theory of relativity provides a structure for space and time with respect to electromagnetic fields. The mathematics defining both special and general relativity is objective and clear-cut. A layman's introduction to this mathematics is presented by Maudlin.[4]

The most famous confirmation of the general theory of relativity was Sir Arthur Eddington's measurement of the deflection of starlight passing near the sun during the 1919 total solar eclipse of the sun. Classical physics following Newton's laws predicted that the starlight would be deflected 0.87 arc seconds; Einstein's theory predicted in would be deflected twice the Newtonian value: 1.75 arc seconds.[5]

Eddington's measurements agreed with Einstein's prediction, not with Newton's. Major newspapers around the world heralded this groundbreaking news on their front pages. When Einstein was asked what his reaction would have been if general relativity had

---

4   Maudlin, *Philosophy of Physics*, 121.

5   Clifford M. Will, "The Confrontation between General Relativity and Experiment," *Living Reviews in Relativity* 17 (June 2014): 4, https://doi.org/10.12942/lrr-2014-4. Another well-known confirmation of general relativity is the precession of the perihelion of Mercury. Classical mechanics underestimates this precession by 7.5 percent. Astronomers in the early twentieth century searched in vain for a planet they dubbed Vulcan, closer to the sun than Mercury, to account for this discrepancy. General relativity provides the correct result within observational error. G. M. Clemence, "The Relativity Effect in Planetary Motions," *Reviews of Modern Physics* 19, no. 4 (October 1947): 361–64, https://doi.org/10.1103/RevModPhys.19.361.

not been confirmed, Einstein famously quipped, "Then I would feel sorry for the dear Lord. The theory is correct anyway."[6]

While General Relativity is often thought to be the province of exploding stars and colliding galaxies, the Global Positioning System (GPS) employs relativity in our daily lives. Clocks provide the bridge between space and time; GPS clocks locate where you are.[7]

GPS satellites orbit approximately 22,200 kilometers (12,600 miles) above the earth's surface. The orbital period at this height is twelve hours; each satellite passes over the same two spots on the earth's equator twice a day.[8] At this altitude, the orbital velocity is approximately 14,000 kph (8,700 mph). Several satellites are always overhead; six to eight satellites are overhead in a 24-satellite system. Position is determined by computing the differences in signal arrival times from four or more satellites. Light travels at 300,000 kilometers per second. At this speed, it takes only 0.000,000,001 of a second for light to travel 30 centimeters (1 foot). Using the time differences from four satellites at a time, current GPS systems in cell phones can locate an observer's position with a 30 centimeter

6   Ilse Rosenthal-Schneider, *Reality and Scientific Truth: Discussions with Einstein, von Laue, and Planck* (Wayne State University Press, 1981), 74. See also Alice Calaprice, ed., *The New Quotable Einstein* (Princeton University Press, 2005), 227.

7   Friedwardt Winterberg proposed putting atomic clocks into space in 1955. He wanted to confirm gravitational time dilation as predicted by Einstein's theory of general relativity. Artificial satellites did not exist at the time of Winterberg's proposal. Winterberg would no doubt be astounded to learn that atomic clocks are precise enough today to measure time dilation with a 0.2 millimeter height difference rather than orbital heights. Friedwardt Winterberg, "Relativistische Zeitdilatation eines künstlichen Satelliten," *Astronautica Acta* 2, no. 1 (1956): 25–29.

8   Holli Riebeek, "Catalog of Earth Satellite Orbits," NASA Earth Observatory, September 4, 2009, https://earthobservatory.nasa.gov/features/OrbitsCatalog/page1.php.

(1 foot) accuracy.[9] Land-surveying tools and other high-end devices employ additional satellite signals and can achieve an accuracy of 2 centimeters.[10]

> The clocks on the satellites appear to run thirty-eight microseconds faster per day than the clocks on Earth.

This is due to a combination of special and general relativity, with the increased clock speed predicted by General Relativity from less gravity overwhelming the time dilation because of the relative velocities of the satellites predicted by special relativity. GPS corrects for this difference. Without this correction, GPS-calculated positions would accumulate up to 10 kilometers error per day (6 miles/day).[11]

GPS satellites confirm Einstein's general theory of relativity: Clocks do indeed run faster in the reduced gravitational field above the surface of the earth. Without correcting for the time contraction due to the decreased gravity at orbital attitude, the GPS system would be miles off.

To the skeptic, Einstein's theory is just hocus-pocus. The skeptic may use GPS systems to keep track of where he is in space and time but maintains that space and time do not exist. He benefits by using twenty-first-century technology, but holds Aristotle's fourth-century BCE view of physics. Yet Einstein's theory is correct, as the

9   Jacob Kastrenakes, "GPS Will Be Accurate Within One Foot in Some Phones Next Year," The Verge, September 24, 2017, https://www.theverge.com/circuitbreaker/2017/9/25/16362296/gps-accuracy-improving-one-foot-broadcom; and Samuel K. Moore, "Superaccurate GPS Chips Coming to Smartphones in 2018," IEEE Spectrum, September 21, 2017, https://spectrum.ieee.org/superaccurate-gps-chips-coming-to-smartphones-in-2018.

10   "GPS Accuracy," GPS.gov, modified March 3, 2022, accessed December 11, 2024, https://www.gps.gov/systems/gps/performance/accuracy/.

11   Winterberg, "Relativistische Zeitdilatation eines künstlichen Satelliten."

evidence proves: The coupled nature of space and time is real. There is no escape.

THE ANSWER

Is gravity an illusion? Yes. Gravity disappears in the general theory of relativity. It is replaced by the curvature of space-time. The general theory of relativity has been proved by countless experiments. One application used by people every day is determining their locations using a GPS. Without adjusting time according to both the relative velocity and the relative gravitational field of the satellites providing the data, the positions computed by using GPS clocks would be miles off the correct locations.

HOW OLD IS THE UNIVERSE?

TO INFINITY AND BEYOND

n 1922, seven years after Einstein published the general theory of relativity, the Russian physicist Alexander Friedmann used Einstein's theory to study a spatially homogeneous and isotropic universe.[1] Friedmann showed the universe described by the general theory was not stable: The general theory implied that the universe was expanding.

Five years later, the Belgian priest Georges Lemaître arrived at the same result and went further: Lemaître showed that, according to General Relativity, galaxies should be receding from Earth

---

1   Friedmann's analysis of a homogeneous isotropic universe is a first order approximation only. The expansion inherent in this first order approximation is unaltered by the universe's observed lumpiness.

at speeds proportional to their distance away from Earth. Lemaître published this result, now known as Hubble's law, two years before Hubble published his landmark paper.[2]

Einstein rejected Friedmann's and Lemaître's analyses. In correspondence with Friedmann, Einstein argued for a static eternal universe. Einstein told Lemaître, "Your calculations are correct, but your physics is atrocious."[3] Einstein had introduced a "cosmological constant" in his 1917 paper to counterbalance the effect of gravity and achieve a static universe.[4] Einstein abandoned the constant in 1931 after Edwin Hubble determined the universe is indeed expanding. Physicist George Gamow reported that Einstein said the cosmological constant was his "biggest blunder."[5]

> The Big Bang theory was born—the idea that the universe emerged from a "primeval atom," a tiny initial state of high density and temperature.

Fred Hoyle (the person who developed the theory of stellar nucleosynthesis that explains the creation of chemical elements by nuclear fusion reactions within stars) coined the phrase *Big Bang* as a derogatory term. Hoyle, along with Einstein pre-1931, was an

2   Georges Lemaître, "Un Univers homogène de masse constante et de rayon croissant rendant compte de la vitesse radiale des nébuleuses extragalactiques," *Annales de la Société Scientifique de Bruxelles* 47 (1927): 49–59.

3   Andre Deprit, "Monsignor Georges Lemaître," in *The Big Bang and Georges Lemaître*, ed. A. Berger (Reidel, 1984), 370.

4   S. E. Rugh and H. Zinkernagel, "The Quantum Vacuum and the Cosmological Constant Problem," *Studies in History and Philosophy of Science Part B: Studies in History and Philosophy of Modern Physics* 33, no. 4 (December 2002): 663–705, https://doi.org/10.1016/S1355-2198(02)00033-3.

5   George Gamow, "The Evolutionary Universe," *Scientific American* 195, no. 3 (September 1956): 136–56, https://doi.org/10.1038/scientificamerican0956-136.

advocate of a static universe, a universe that is neither expanding nor contracting, a universe that stays the same forever.[6] It would take measurements of both the velocities of the galaxies in the universe and of the remnant radiation left over from the Big Bang to establish that the universe is indeed expanding.

By "the universe is expanding," physicists mean space itself is expanding. Since the law of conservation of mass-energy states that the total mass-energy of the universe is a constant, this means that the density of mass-energy in the early universe was much higher than it is today. It is a mistake to think of the Big Bang occurring at a spot in infinite space with galaxies flying apart in preexisting space. Rather, it is better to think of this as all space filled with an extremely high-density plasma throughout space in the early universe, and a decrease in the average mass-energy density as space expands. With the continued expansion of space into the future, the average mass-energy density of the universe will eventually approach zero.

Measuring the universe is a difficult enterprise. The distances to nearby stars can be measured by using parallax. As illustrated in Figure 1, the positions of nearby stars shift slightly against the background of faraway stars as the earth orbits the sun. Thus, nearby stars can be triangulated, using the diameter of earth's orbit as the

---

6   Stellar nucleosynthesis was first explained by Fred Hoyle in 1946. However, the Nobel Prize for developing the theory of stellar nucleosynthesis was awarded to William A. Fowler in 1983 for a 1957 paper coauthored by Margaret Burbidge, Geoffrey Burbidge, William A. Fowler, and Fred Hoyle. Fowler, in his own Nobel lecture, wrote: "The grand concept of nucleosynthesis in stars was first definitely established by Hoyle in 1946." Hoyle's mistaken views about the Big Bang are thought to have eliminated him from receiving the Nobel Prize. *Nature* editor John Maddox called it "shameful" that Fowler had been rewarded with a Nobel Prize and Hoyle had not. John Maddox, "Obituary: Fred Hoyle (1915–2001)," *Nature* 413, no. 6853 (September 2001): 270, https://doi.org/10.1038/35095162.

narrow side of a very long, thin triangle. With current precision, this approach can measure the distances to stars up to 16,000 light-years.[7]

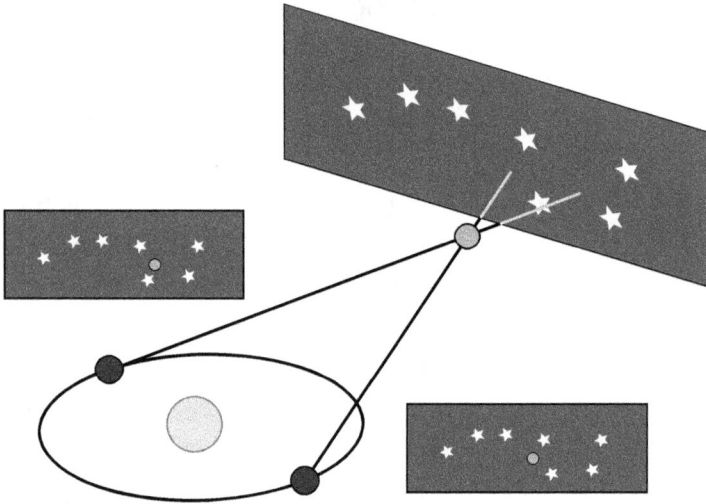

Figure 1. The positions of nearby stars appear to move against the fixed background of stars very far from the earth.[8]

Luminosity can determine distance to stellular objects too far away to be resolved by using solar parallax and triangulation. Astronomers look at nearby stars, categorize them by type, compute their distances, determine their relative luminosities, and endeavor to find a relationship between distance and apparent brightness. The "first inch" in this yardstick was the work of Henrietta Leavitt, who discovered a relationship between the luminosity of variable

7   Adam G. Riess et al., "Parallax Beyond a Kiloparsec from Spatially Scanning the Wide Field Camera 3 on the Hubble Space Telescope," *The Astrophysical Journal* 785, no. 2 (April 2014): 161, https://doi.org/10.1088/0004-637X/785/2/161.
8   Image by Kes47 and WikiStefan (Wikimedia Commons, CC BY-SA 3.0).

Cepheid stars and their period of oscillation. Leavitt catalogued 1,777 Cepheid variables working as a "computer" at Harvard College Observatory in 1908.[9] The relationship Leavitt discovered is now known as Leavitt's law.

Astronomers glean a mountain of information from a speck of light. They do this by a procedure called *spectrometry*.

In 1814, the German physicist Joseph von Fraunhofer was studying the spectrum of the sun. He noticed a series of dark lines in this spectrum, as presented in Figure 2. Two centuries earlier, Isaac Newton had used a prism to decompose sunlight into the rainbow of colors displayed in this figure, but the quality of Newton's prism was too poor to show the dark lines.[10] Fraunhofer could discern 574 dark fixed lines in the solar spectrum. Today, scientists have identified millions of these lines.[11]

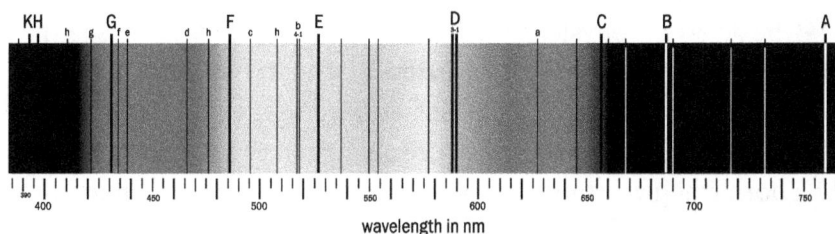

Figure 2. Light from the sun decomposed into a spectrum of colors showing the appearance of dark "Fraunhofer" lines.

9   Henrietta S. Leavitt, "1777 variables in the Magellanic Clouds," *Annals of Harvard College Observatory* 60 (1907): 87–108.

10   Newton captured the rainbow of colors produced by the first prism using a second prism. Properly adjusted, the light coming out of the second prism is white. In this way, Newton showed that white light is a mixture of colors. Isaac Newton, *Opticks: Or A Treatise of the Reflections, Refractions, Inflections and Colours of Light* (Royal Society, 1705), 13–19.

11   Myles W. Jackson, *Spectrum of Belief: Joseph Von Fraunhofer and the Craft of Precision Optics* (MIT Press, 2000), 1–16.

Forty-five years later, Gustav Kirchhoff noticed that some of the Fraunhofer lines coincided with the bright lines emitted by heating various chemical elements.[12] He concluded, correctly, that the absorption of light caused the dark Fraunhofer lines in the solar spectra at these frequencies by the chemical elements he had been heating.[13]

Spectral lines are unique to each chemical element and can identify elements in the same way as fingerprints are used to identify people. The dark lines in the solar spectra reveal which elements are found in the sun's corona. Helium was detected in the sun's corona before it was found on earth. The English chemist Norman Lockyer introduced the word "helium," from the Greek word for the sun, "helios," in 1868, for the element producing distinct spectral lines but not yet detected on earth.[14] Twenty-seven years later, another English chemist, Sir William Ramsay, isolated a sample of helium on earth by treating the mineral uraninite with acids.[15]

The Austrian physicist Christian Doppler proposed in 1842 that the observed frequency of a wave depends on the velocity of a source relative to an observer. We are accustomed to the Doppler effect, as this principle is now known, by the change in the pitch of a whistle or a horn when we stand near a fast-moving train or car. There were

12 Gustav Kirchhoff, "Ueber die Fraunhofer'schen Linien" (On Fraunhofer's lines), lecture, October 20, 1859, printed in *Monatsbericht der Königlichen Preussische Akademie der Wissenschaften zu Berlin: Aus dem Jahre 1859* (Königlichen Akademie der Wissenschafter, 1860), 662–65.

13 Gustav Kirchhoff, "Ueber das Verhältniss zwischen dem Emissionsvermögen und dem Absorptionsvermögen der Körper für Wärme und Licht," *Annalen der Physik* 185, no. 2 (1860): 275–301, https://doi.org/10.1002/andp.18601850205.

14 G. A. Wilkins, "Sir Norman Lockyer's Contributions to Science," *Quarterly Journal of the Royal Astronomical Society* 35 (1994): 51–57.

15 William Ramsay, "On a Gas Showing the Spectrum of Helium, the Reputed Cause of D_3, One of the Lines in the Coronal Spectrum. Preliminary Note, *Proceedings of the Royal Society of London* 58, no. 347–352 (January 1895): 65–67, https://doi.org/10.1098/rspl.1895.0006.

no fast-moving vehicles in Doppler's day, so Doppler applied his principle to the study of the colors of binary stars.[16] His effort in this regard was misplaced because the colors of the stars are primarily determined by their temperatures, but, in 1868, the British astronomer William Huggins was the first to determine the velocity of a star relative to the earth by noting the shift in the star's Fraunhofer lines.[17]

If a star is moving away from the earth, the star's Fraunhofer lines will shift toward the red end of the spectrum; the lines will shift toward the blue if the star is moving toward the earth. The redshift of starlight is illustrated in Figure 3. The colors on both spectra in this illustration are the same; it is the locations of the Fraunhofer lines that have shifted.

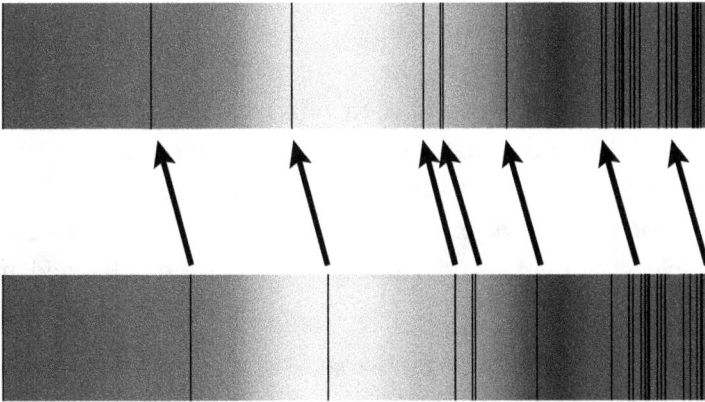

Figure 3. The colors remain the same, but the Fraunhofer lines shift toward the red end of the spectrum for a light-emitting object moving away from the observer.[18]

16   Alec Eden, *The Search for Christian Doppler* (Springer, 2012).
17   William Huggins, "Further Observations on the Spectra of Some of the Stars and Nebulae, with an Attempt to Determine Therefrom Whether These Bodies are Moving towards or from the Earth, Also Observations on the Spectra of the Sun and of Comet II," *Philosophical Transactions of the Royal Society of London* 158 (January 1868): 529–64, https://doi.org/10.1098/rstl.1868.0022.
18   Image by Georg Wiora (Dr. Schorsch) (Wikimedia Commons, CC BY-SA 3.0).

The existence of galaxies outside of the Milky Way was not established until the twentieth century. Prior to the work of Vesto Slipher, director of the Lowell Observatory in Flagstaff, Arizona, the smudges of light visible through telescopes and recorded in photographs were called nebulae and were thought to be located within the Milky Way. Beginning with observations in 1912, Slipher showed that most spiral nebulae, now called galaxies, exhibited significant redshifts of the Fraunhofer lines in their spectrums.[19] These redshifts showed that these nebulae (galaxies) are receding away from the earth at great velocities. An exception to the receding galaxies was Andromeda spiral which is approaching the earth at 300 km/s. At this velocity, the Andromeda galaxy will collide with the Milky Way in four billion years.[20]

Edwin Hubble used Leavitt's law to compute the distances to galaxies and combined it with Slipher's redshift approach to compute relative speeds of galaxies. The result was a relationship between the distances to galaxies and their relative motion with respect to us. Hubble's law states that the relative speed of separation between galaxies is proportional to the distance between them.[21] Hubble's estimate of the rate of expansion was low by a factor of seven.[22] Nonetheless, Slipher's and Hubble's observations changed man's

19   Vesto Slipher, "Spectrographic Observations of Nebulae," *Popular Astronomy* 23 (January 1915): 21–24.

20   Riccardo Schiavi et al., "Future Merger of the Milky Way with the Andromeda Galaxy and the Fate of Their Supermassive Black Holes," *Astronomy & Astrophysics* 642 (October 2020): A30, https://doi.org/10.1051/0004-6361/202038674.

21   Edwin Hubble, "A Relation Between Distance and Radial Velocity Among Extra-Galactic Nebulae," *Proceedings of the National Academy of Sciences* 15, no. 3 (April 1929): 168–73, https://doi.org/10.1073/pnas.15.3.168.

22   Adam G. Riess et al., "Milky Way Cepheid Standards for Measuring Cosmic Distances and Application to Gaia DR2: Implications for the Hubble Constant," *The Astrophysical Journal* 861, no. 2 (July 2018): 126, https://doi.org/10.3847/1538-4357/aac82e.

view of existence from a static universe with everything contained in the Milky Way galaxy to an expanding universe containing billions of galaxies like the Milky Way.

Friedmann's and Lemaître's predictions that the universe is expanding, followed by Hubble's measured confirmation of an expanding universe, appears to be a match made in heaven. Georges Lemaître, a Catholic priest and future president of the Pontifical Academy of Sciences, certainly thought so. In a paper published in 1931, Lemaître projected the universe back in time and concluded that the entire mass of the universe was concentrated into a single point at a finite time in the past. This "primeval atom" was "the beginning of the world," Lemaître wrote, where and when the fabric of time and space came into existence. One hears echoes of the Biblical creation story in Lemaître's concept of the universe starting with a Big Bang:

> If the world has begun with a single *quantum*, the notions of space and time would altogether fail to have any meaning at the beginning; they would only begin to have a sensible meaning when the original quantum had been divided into a sufficient number of quanta. If this suggestion is correct, the beginning of the world happened a little before the beginning of space and time.[23]

Of this concept, Maudlin writes:

> The notion that space-time itself may have a beginning or an end is the most philosophically intriguing implication of General Relativity, but also the most tenuous.[24]

---

23 Georges Lemaître, "The Beginning of the World from the Point of View of Quantum Theory," *Nature* 127, no. 3210 (May 1931): 706, https://doi.org/10.1038/127706b0.
24 Maudlin, *Philosophy of Physics*, 146.

It is tenuous on multiple levels. Projecting the equations of General Relativity back in time generates a state of infinite density. Mathematical expressions that produce infinite values are called singularities, and singular values are meaningless. Something is missing from the model. A correct model of the early universe would give meaningful results. Despite the infinity in his result, Lemaître says space and time did not exist if you go back in time beyond infinite density. The leap from "I have a mathematical expression that goes to infinity" to "There is no time and space beyond the mathematical infinity I found" is arbitrary.

No evidence exists for this assertion and thus must be dismissed out of hand.

THE ANSWER

How old is the universe? The consensus age of the universe is 13.787 billion years plus or minus twenty million years.[25] This value originates with Hubble's law, which states that the further a galaxy is from the Milky Way (our galaxy), the faster the galaxy is receding away from us. Extrapolating these galaxy motions back in time indicate that the universe was much smaller and denser in the early universe than it is today. The beginning of the universe is called the Big Bang because of the explosive growth of the universe. However, the equations describing the beginning of the universe go to infinity as time approaches zero. Since infinite values are meaningless, no information is available on what happened during the Big Bang or before the Big Bang.

---

25 Planck Collaboration, "Planck 2018 Results: VI. Cosmological Parameters," *Astronomy & Astrophysics* 641 (September 2020): A6, https://doi.org/10.1051/0004-6361/201833910.

DID THE BIG BANG REALLY HAPPEN?

THE EXPANDING UNIVERSE

n 1823, German astronomer Heinrich Olbers was puzzled why the night sky is dark.

> If the universe is infinite and static, populated by an infinite number of stars, any line of sight from Earth must end at the surface of a star.

Hence, Olbers reasoned, the night sky should be as bright as the surface of the sun.

Edgar Allan Poe suggested a solution to Olbers's paradox in his prose poem *Eureka*.[1] Poe's solution was to say that the age of the

---

1   Edgar Allan Poe, *Eureka: A Prose Poem* (Putnam, 1848).

universe was finite and "the distance of the invisible background so immense that no ray from it has yet been able to reach us at all." Except calculations show that a finite, static universe would heat the hydrogen plasma filling the universe to 3,000 K, rendering outer space opaque.[2] The sky is five hundred billion times darker than it would be if the universe was too young to have reached equilibrium.

The only known explanation for the dark night sky is the expanding universe. With galaxies retreating from each other at greater and greater speeds the further they are apart, light is shifted to lower frequencies until it disappears entirely. Without the Big Bang and the subsequent expansion of the universe, the night sky would be aglow and life on earth would be impossible.

Yet, our understanding of how the universe evolved involves contradictions. Special relativity is incompatible with general relativity, gravity does not work as expected, the universe is expanding too fast, the level of cosmic microwave background radiation is too uniform, and observations using the James Webb Space Telescope reveal that stars and galaxies formed much earlier than was thought possible. Each of these issues is examined below.

The Lorentz transformation of special relativity assumes a "flat" space, whereas the space of general relativity is inherently warped.

Special relativity applies only in regions of constant gravity, while general relativity is entirely about the bending of space. Both theories give beautiful, amazing, and meaningful results, but it is a mistake to think these two contradictory equations capture everything that exists.

---

2   Mark Kidger, *Cosmological Enigmas: Pulsars, Quasars, and Other Deep-Space Questions* (Johns Hopkins University Press, 2007), 144–45.

Gravity does not work as expected. In the solar system, the gravitational field decreases as you go away from the sun. As a result, planets further from the sun orbit at slower velocities. However, galaxies behave differently. Even accounting for the different distribution of matter in a galaxy compared to the distribution of matter in the solar system, stars in spiral galaxies rotate much faster than predicted by the equations of general relativity as you go outward from the center of the galaxy. This is illustrated in Figure 1.

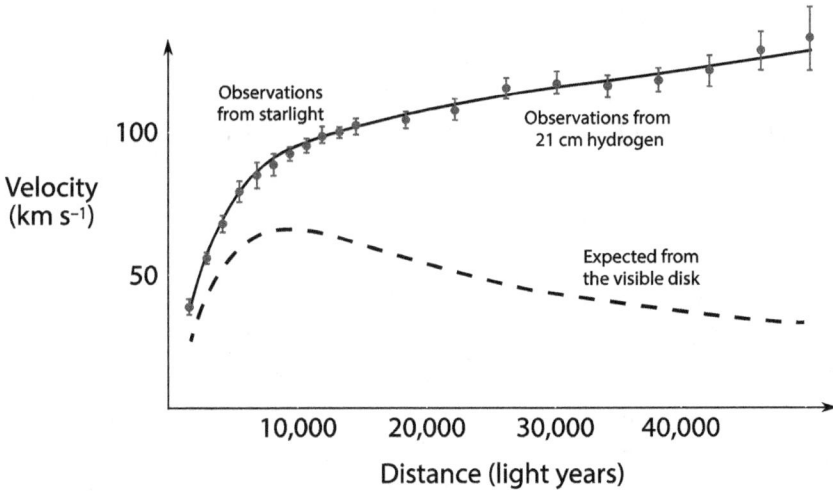

Figure 1. The rotation velocity of stars and gas as a function of their distance from the center of a galaxy. The dashed curve indicates the velocities of the stars rotating around their common center if gravity alone from the visible mass of stars and gas held them together. The solid line provides the rotational velocity curve obtained from the measured data.[3]

Three possible explanations of the anomalous rotational velocities of stars in galaxies are:

---

3   Mario De Leo (Wikimedia Commons, CC BY-SA 4.0).

1. The general theory of relativity is wrong.
2. A form of matter, called *dark matter*, exists that provides gravity but otherwise does not interact with ordinary matter.
3. Unaccounted forces exist that hold the fast-moving stars in their orbits.

Alternative theories to general relativity abound, but are unappealing.[4] For this reason, physicists have largely embraced "dark matter" to explain the anomalous rotational velocity of galaxies. Dark matter is a hypothetical form of matter, not observed on earth or even in the solar system. It makes its presence known only at galactic scales through its interaction with gravity. To explain the velocity curve in Figure 1, dark matter must outweigh ordinary matter by six to one.[5]

Not only do stars in galaxies rotate too fast, but the expansion rate of the universe is also too fast. In 1999, observations of Type Ia supernova in distant galaxies, along with their redshifts, showed that the expansion of the universe is accelerating.[6] Saul Perlmutter, Brian Schmidt, and Adam Riess received the 2011 Nobel Prize in Physics for this work. Einstein's cosmological constant reappeared, this time not to hold the universe in a steady state, but to accelerate it beyond the Friedmann–Lemaître model. The energy required to provide this acceleration is called *dark energy* because no one knows from whence it comes. Dark energy contributes 68 percent of the total

4 Philip D. Mannheim, "Alternatives to Dark Matter and Dark Energy," *Progress in Particle and Nuclear Physics* 56, no. 2 (April 2006): 340–445, https://doi.org/10.1016/j.ppnp.2005.08.001.

5 "Dark Matter," CERN, accessed February 1, 2025, https://home.cern/science/physics/dark-matter.

6 S. Perlmutter et al., "Measurements of Omega and Lambda from 42 High-Redshift Supernovae," *Astrophysical Journal* 51, no. 2 (1999): 565–86, https://doi.org/10.1086/307221.

energy in the present-day observable universe, dark matter contributes 27 percent, while ordinary matter contributes only 5 percent.[7]

Adding dark matter and dark energy to the Friedmann–Lemaître model brings it in line with the universe's observed expansion rate, but it still fails to explain the observed level and uniformity of the cosmic microwave background (CMB) radiation. CMB was discovered by Bell Telephone Laboratories engineers Arno Penzias and Robert Wilson in 1964. Working to develop a satellite communication system in Holmdel, New Jersey, they could not explain an ever-present, unidirectional microwave signal in their equipment.[8] Penzias called Robert Dicke, professor of physics at Princeton University, who explained that the microwave signal was a remnant of the Big Bang. In answering *Question 29—How old is the universe?*, the Big Bang is an expansion of space with a constant total mass-energy in the universe through time. At the beginning of the universe, the energy density of the universe was extremely high and the corresponding radiation was extremely hot. As the universe expanded, the radiation redshifted, causing the background radiation to cool. The CMB measured today corresponds to a radiation temperature of 2.725 K or −270.425°C.[9]

To avoid conflict, Penzias, Wilson, and Dicke agreed to publish papers simultaneously in the *Astrophysical Journal Letters*. One

---

7   Ruth Durrer, "What Do We Really Know About Dark Energy?," *Philosophical Transactions of the Royal Society A: Mathematical, Physical and Engineering Sciences* 369, no. 1957 (December 2011): 5102–114, https://doi.org/10.1098/rsta.2011.0285.

8   K is the symbol for Kelvin, the base unit for temperature in the International System of Units. It uses the same scale as Celsius degrees, shifted to absolute zero, −273.15°C.

9   P. Noterdaeme et al., "The Evolution of the Cosmic Microwave Background Temperature: Measurements of T_{CMB} at High Redshift from Carbon Monoxide Excitation," *Astronomy and Astrophysics* 526 (February 2011): L7, https://doi.org/10.1051/0004-6361/201016140.

paper, the one by Penzias and Wilson, described the experimental results;[10] the other paper, the one by Dicke, explained the source of the radiation.[11] Oddly, while Penzias and Wilson received the 1978 Nobel Prize for Physics for their experimental work, Dicke, the person who explained the origin of the CMB radiation, was snubbed by the Nobel Committee.

However, postulating the cosmic microwave background radiation to be a remnant of the Big Bang theory does not match observations. The CMB is too smooth by many orders of magnitude to explain the early evolution of the universe. To bring the Big Bang theory in line with observed data, in 1981, Alan Guth proposed that the early universe underwent a period of rapid inflation. Cosmological inflation lasted from 10^{-36} seconds after the Big Bang to between 10^{-33} and 10^{-32} seconds after the Big Bang singularity.[12] To ensure that the Universe appears flat, homogeneous, and isotropic, it is necessary for the Universe to have expanded by a factor of at least 10^{26} during inflation.[13]

The accepted model of the universe today contains three "fudge factors using four variables."

Dark matter has never been observed in our solar system, dark energy is created out of nothing, and inflation is an ad hoc device

10   A. A. Penzias and R. W. Wilson, "A Measurement of Excess Antenna Temperature At 4080 Mc/s," *Astrophysical Journal Letters* 142 (July 1965): 419–21, https://doi.org/10.1086/148307.

11   R. H. Dicke et al., "Cosmic Black-Body Radiation," *Astrophysical Journal Letters* 142 (July 1965): 414–19, https://doi.org/10.1086/148306.

12   Alan H. Guth, "Inflationary Universe: A possible Solution to the Horizon and Flatness Problems," *Physical Review D* 23, no. 2 (January 1981): 347–56, https://doi.org/10.1103/PhysRevD.23.347.

13   Andrei Linde, *Particle Physics and Inflationary Cosmology* (CRC Press, 1990).

to make the numbers come out right.[14] And, recently, observations of early universe using the James Webb Space Telescope (JWST) have revealed that stars and galaxies developed much earlier than thought possible:

This discovery proves that luminous galaxies were already in place 300 million years after the Big Bang and are more common than what was expected before JWST… Galaxy formation models will need to address the existence of such large and luminous galaxies so early in cosmic history.[15]

Is it possible that cosmologists have missed something in their theories?

Nobel laureate Hannes Alfvén suggested in the late twentieth century that electric and magnetic forces shape galactic structures along with gravity.[16] Alfvén was an expert on plasmas, a gaseous form of matter containing ions. Our primary experience with plasmas on earth is as lightning strikes.[17] However, almost all matter in

---

14   Paul J. Steinhardt, "The Inflation Debate: Is the Theory at the Heart of Modern Cosmology Deeply Flawed?," *Scientific American* 304, no. 4 (April 2001): 18–25, https://doi.org/10.1038/scientificamerican0411-36.

15   Stefano Carniani et al., "Spectroscopic Confirmation of Two Luminous Galaxies at a Redshift of 14," *Nature* 633 (July 2024): 318–22, https://doi.org/10.1038/s41586-024-07860-9.

16   H. O. G. Alfven, "Cosmology in the Plasma Universe: An Introductory Exposition," *IEEE Transactions on Plasma Science* 18 (February 1990): 5–10, https://doi.org/10.1109/27.45495.

17   In a thunderstorm, rising water droplets and tiny ice crystals rub against falling, larger hail pellets, causing the rising droplets to become positively charged and the falling pellets to become negatively charged. The voltage difference between the upper section of the thunderstorm cloud and the middle and lower sections rises until the air ionizes, releasing a stroke of lightning. Electrons are stripped from the molecules of the heated air forming a plasma. Although the breakdown of the air begins in the upper reaches of…

the universe exists in plasma form: stars are composed of dense plasmas, while interstellar and intergalactic space form very large, very low-density plasmas.

Plasma is complicated and chaotic. The timing and location of a lightning strike is unpredictable.

Alfvén's work has been largely ignored by cosmologists for several reasons, including Alfvén's view that the Big Bang never happened, that he published primarily in electrical engineering journals, and because plasma physics is complicated and chaotic.[18] A recent paper summarized the situation as follows:

> ...most cosmologists hardly think about magnetism. "Everyone knows it's one of those big puzzles," [cosmologist Levon Pogosian, professor of physics at Simon Fraser University, BC, Canada] said. But for decades, there was no way to tell whether magnetism is truly ubiquitous and thus a primordial component of the cosmos, so cosmologists largely stopped paying attention.[19]

However, magnetic fields of galactic scope have recently been observed. In 2019, Federica Govoni of the National Institute for Astrophysics in Cagliari, Italy, along with twenty-seven coauthors, detected a magnetic filament ten million light-years long spanning

...the cloud, electrons flow upward, from the ground toward the top of the cloud, in this process. The current flow in a lightning strike generates magnetic fields with sufficient power to disrupt communications and electric power distribution facilities.

18 H. O. G. Alfvén, "Cosmology: Myth or Science?," *IEEE Transactions on Plasma Science* 20 no. 6 (December 1992): 590–600, https://doi.org/10.1109/27.199498; and Helge Kragh, *Cosmology and Controversy: The Historical Development of Two Theories of the Universe* (Princeton University Press, 1996), 482–83.

19 Natalie Wolchover, "The Hidden Magnetic Universe Begins to Come Into View," *Quanta Magazine*, July 2, 2020, https://www.quantamagazine.org/the-hidden-magnetic-universe-begins-to-come-into-view-20200702/.

the space between two clusters of galaxies.[20] The Milky Way is approximately one hundred thousand light-years across, so this filament is one hundred times the diameter of the Milky Way. Ampère's law states that magnetic fields are generated by currents. For this magnetic field to exist, the plasma within a galaxy and between galaxies must support the flow of ions. The forces produced by electric currents and magnetic fields of this size will affect the shapes of galactic structures. Further, in a recent paper, Karsten Jedamzik and Levon Pogosian show that including magnetic fields in simulations alters the value of the Hubble constant.[21] Interstellar and intergalactic plasma form a significant component of the universe, after all.

THE ANSWER

Did the Big Bang really happen? Yes, the evidence for the Big Bang is overwhelming. However, the present models of the universe include ad hoc factors such as cosmic inflation, dark matter, and dark energy that raise questions about current cosmology models. Future scientists will no doubt provide better theories of the universe's evolution.

20 Federica Govoni et al., "A Radio Ridge Connecting Two Galaxy Clusters in a Filament of the Cosmic Web," *Science* 364, no. 6444 (June 2019): 981–84, https://doi.org/10.1126/science.aat7500.

21 Karsten Jedamzik and Levon Pogosian, "Relieving the Hubble Tension with Primordial Magnetic Fields," *Physical Review Letters* 125 (October 2020), https://doi.org/10.1103/PhysRevLett.125.181302.

HOW DOES MY BRAIN WORK?

THE PHYSICS OF THE BRAIN

For fifteen years, Cathy Hutchinson remained motionless and nonverbal in a wheelchair. At forty, she had experienced paralysis from the neck down due to a severe stroke while gardening, severing the connections between her brain and limbs. Then a miracle happened.[1] On one particular day, Ms. Hutchinson lifted a bottle to her mouth and sipped coffee through a straw for the first time in a decade and a half. She used a metallic blue robotic arm, which responded to instructions from a microelectrode array implanted into her brain. This array was placed in Ms. Hutchinson's

---

1   "Miracle" is used here in the secular sense: a wonderful, life-changing event that occurs as a consequence of extraordinary human effort and achievement.

motor cortex, the region responsible for arm movement. Before her paralysis, she had sent signals to her arm muscles countless times to raise her arms. Now, with electrodes in her brain, the command to raise her arm was transmitted from her brain to a computer equipped with decoding software for her neural signals. The robotic arm, in tandem with the computer, was programmed to respond as if it were a natural human limb.[2]

The video footage capturing this momentous occasion depicts Ms. Hutchinson immersed in deep concentration, her undivided attention fixed on the current task. With precision, she haltingly maneuvers the robotic arm to bring the bottle to her lips, taking a sip of coffee through the straw. Setting the bottle aside, a triumphant smile lights up her face, and her eyes exude a sense of achievement. The researchers respond with cheers of celebration.[3]

> That thoughts control motion is uncontroversial.
> We all do it. However, the idea that thoughts
> can manipulate machines *using only the*
> *mind* is on the frontier of science.

Since Ms. Hutchinson's inspiring achievement, many brain/machine interfaces have been developed. Elon Musk formed a company called Neuralink in 2017 to commercialize the technology. The first person to receive a Neuralink implant, Noland Arbaugh, an individual who was paralyzed below the shoulders in a diving accident, is able to move a cursor on a screen using only his thoughts. Noland

2   Leigh R. Hochberg et al., "Reach and Grasp by People with Tetraplegia Using a Neurally Controlled Robotic Arm," *Nature* 485, no. 7398 (May 2012): 372–75, https://doi.org/10.1038/nature11076.

3   Nature Video, "Paralysed Woman Moves Robot with Her Mind—by Nature Video," May 16, 2012, YouTube, 4:29, https://www.youtube.com/watch?v=ogBX18maUiM.

marvels, "Now I'm beating my friends at [video] games, which really shouldn't be possible but it is.[4]

In 1780, Luigi Galvani discovered that electricity causes muscles to twitch. Electricity was all the rage among intellectuals. Pieter van Musschenbroek, physics professor at the University of Leiden, had invented the Leiden jar, a simple device that allowed people to store an electric charge for the first time.[5] A literal "lightning in a bottle," the Leiden jar provided amusements for the upper class, shocking guests with its power and mystery.

Luigi Galvani was dissecting a frog using a scalpel near a Leiden jar when it discharged. The frog's severed leg twitched with the jolt of electricity. Galvani had discovered the electricity-body connection.[6] The mechanism behind the frog's leg movement was initially hotly debated. Alessandro Volta, the inventor of the battery and the man after whom voltage is named, argued that, yes, electricity moves the frog's leg, but he argued that electricity was not responsible for ordinary animal locomotion. Proof that Volta was wrong, that electricity is indeed involved in transmitting signals along nerves, was left to Julius Bernstein at the University of Heidelberg a century and

4   Lara Lewington, Liv McMahon, and Tom Gerken, "The Man with a Mind-Reading Chip in His Brain—Thanks to Elon Musk," BCC, March 22, 2025, https://www.bbc.com/news/articles/cewk49j7j1p0.

5   A Leiden jar is constructed by coating both the inside and the outside of a glass jar with separate conducting tinfoils and attaching an electrode to each foil. The foil-glass-foil arrangement forms a capacitor for the storage of electric charge. Modern materials and manufacturing methods have turned this basic concept into supercapacitors able to provide energy boosts to Formula 1 racing cars, as well as a host of other applications. Cibelle Celestino Silva and Peter Heering, "Re-examining the Early History of the Leiden Jar: Stabilization and Variation in Transforming a Phenomenon into a Fact," *History of Science* 56, no. 3 (April 2018): 314–42, https://doi.org/10.1177/0073275318768418.

6   Guiliano Pancaldi, *Volta: Science and Culture in the Age of Enlightenment* (Princeton University Press, 2003).

a half later.[7] Bernstein was able to isolate a nerve, place it on his apparatus, and precisely record how the nerve current changes over time. Bernstein's measurements recorded what is known today as an *action potential*, a characteristic pattern of changes in the electrical properties of a cell. We now know that action potentials are the principal means of communication between cells. An action potential travels across a cell's membrane like a wave. When an action potential reaches the knobby end of a neuron, it pushes out neurotransmitters. These chemicals reach neighboring cells and trigger action potentials in them as well. With a frog leg, an action potential reaching the muscle tissue causes it to twitch.

Action potential is a bioelectric phenomenon, involving both electrical and chemical processes. To explore the chemical side of action potentials, researchers Alan Hodgkin and Andrew Huxley at the Marine Biological Association of the United Kingdom in Plymouth, England, performed a series of experiments in 1952 involving ion transfer through neuron cell membranes. They placed a squid axion into a liquid bath and measured the action potential as they selectively altered the concentrations of sodium, potassium, and other ions. By controlling the voltage across the cell membrane, Hodgkin and Huxley could change the flow of ions into and out of the cell. The result of all this work was a mathematical model, called the Hodgkin-Huxley equations, that describes how action potentials in neurons are initiated and propagated. Hodgkin and Huxley received the Noble Prize in Physiology or Medicine in 1963 for explaining signal propagation through neurons.

The French physician Louis Lapicque was the first to develop an equivalent circuit to model the voltage versus current relationship

---

7   Ernst-August Seyfarth, "Julius Bernstein (1839–1917): Pioneer Neurobiologist and Biophysicist," *Biological Cybernetics* 94, no. 1 (December 2005): 2–8, https://doi.org/10.1007/s00422-005-0031-y.

in neutrons.[8] Electrical engineers use equivalent circuits to represent individual components in complex electronic devices. A complex electronic device is simulated by combining equivalent circuits of individual components to build a model of the entire device. In Lapicque's case, he created an equivalent circuit for the neuron and showed good agreement between measurements and model predictions for the nerve response time with respect to applied voltage.

A key aspect of neuron operation is the all-or-nothing principle, first identified by the English physiologist Edgar Adrian in the 1920s.

> The all-or-nothing principle states that a neuron either emits an action potential or it does not—nothing in between.

This behavior was captured by the *leaky integrate-and-fire neuron* model developed in the 1960s. This model combines Lapicque's equivalent circuit with the Hodgkin-Huxley equations. The leaky integrate-and-fire neuron model is leaky because some of the current leaks away, and it "integrates" because the model stores the charge not leaked, and it "fires" an action potential when the charge reaches a threshold value. After "firing," the action potential in the neuron returns to its base value.

The human brain contains eighty-five billion neurons, connected by over one hundred trillion synapses. Every second you are alive, billions of action potentials course through your brain.[9]

A neuron provides a logical "1" or a "0," just like a transistor in a computer. The neuron either fires or it does not. Claude Shannon, in

---

8   Nicholas Brunel and Mark C. W. van Rossum, "Lapicque's 1907 Paper: From Frogs to Integrate-and-Fire," *Biological Cybernetics* 97 (October 2007): 337–39, https://doi.org/10.1007/s00422-007-0190-0.

9   The human brain comprises over 3,300 cell types. "Brain Cell Census," special issue, *Science* 382, no. 6667 (October 2023), https://www.science.org/toc/science/382/6667.

the 1940s, showed that all coded information can be represented by a term he called *bits*, logical 1s or 0s. The brain is an information-processing machine, and the "leaky integrate-and-fire neuron" provides the mechanism by which the brain processes information.

Representing bits as the firing of neurons rather than as action potential values is an example of *frequency-based coding*.[10] This type of coding is evident in the visual system, where the firing frequency of the neurons that transmit information is a function of the intensity of light.

> A neuron in a brain provides one bit of information,
> just like a transistor in a computer.

The "brain is a computer" hypothesis is based on this neuron/transistor analogy.[11] Under this hypothesis, the mind/body problem is solved: The mind is information flowing through the brain, analogous to the information flowing through a computer. The physical components in the brain, namely the neurons, "fire" 1s and 0s, analogous to the 1s and 0s in an electronic computer. The pattern of 1s and 0s gives rise to consciousness. The software in a computer is not physical, nor is the consciousness in your brain. According to Grace Lindsay:

> To think of the brain as a computing device following the rules of logic—rather than just a bag of proteins and chemicals—would open the door to understanding thought in terms of neural activity.[12]

---

10   S. J. Thorpe, "Spike Arrival Times: A Highly Efficient Coding Scheme for Neural Networks," in *Parallel Processing in Neural Systems and Computers*, eds. R. Eckmiller et al. (North-Holland, 1990), 91–94.

11   Margaret A. Boden, *Mind as Machine: A History of Cognitive Science* (Clarendon Press, 2006).

12   Grace Lindsay, *Models of the Mind: How Physics, Engineering and Mathematics Have Shaped Our Understanding of the Brain* (Bloomsbury Sigma, 2022), 53.

Except that thought isn't that easy. The Hodgkin-Huxley equations are nonlinear, and nonlinear equations provide chaos. Digital computers are deterministic and predictable. Chaotic systems are unpredictable and are characterized by attractor states. Computers are the opposite: Engineers work exceedingly hard to make computers do exactly what they are told. Any difference between expected computer output and actual performance is called a *bug* and is squashed. Digital computers are predictable; biological networks are chaotic.

In 1996, Van Vreeswijk and Sompolinsky ran simulations on clusters of neurons and discovered that changing the starting state of a single neuron—from firing to not firing, or vice versa—created distinct patterns of action potentials across the neurons.[13] The result was the paper "Chaos in Neuronal Networks with Balanced Excitatory and Inhibitory Activity."[14] "The brain is a digital computer" changed into "the brain is a chaotic computer." A thought may shift a human brain from one attractor state into another. Physics and common experience are the same: People change the state of their neural circuitry by what they think.

THE ANSWER

How does my brain work? The brain is a neural network with eighty-five billion neurons connected by over one hundred trillion synapses. The various functions of the brain—motion, memory, vision, speech, and reasoning—employ different neural structures. All neurons, however, employ the "all-or-nothing" principle providing logical 1s and 0s, analogous to the 1s and 0s in digital computers.

---

13   Charles F. Stevens and Anthony M. Zador, "Input Synchrony and the Irregular Firing of Cortical Neurons," *Nature Neuroscience* 1, no. 3 (1998): 210–17, https://doi.org/10.1038/659.

14   Jonathan Kadmon and Haim Sompolinsky, "Transition to Chaos in Random Neuronal Networks," *Physical Review X* 5 (November 2015): 041030, https://doi.org/10.1103/PhysRevX.5.041030.

HOW ARE MEMORIES STORED?

THE HOPFIELD NETWORK

M emory is contained in neural circuitry, an inherently chaotic system. When we search for a memory, it takes time to access the attractor state. Further, the memory may not be an exact copy of the way we experienced it. Thus, our memory recall may be in the correct attractor region, but not spot-on. Also, similar memories in the same attractor region may be difficult to separate and merge. This is characteristic of chaotic systems.

A network model of memory was published by John J. Hopfield in 1982.[1] Hopfield's paper, "Neural Networks and Physical Systems

---

1   J. J. Hopfield, "Neural Networks and Physical Systems with Emergent Collective Computational Abilities," *Proceedings of the National Academy of Sciences* 79, no. 8 (April 1982): 2554–58, https://doi.org/10.1073/pnas.79.8.2554.

with Emergent Collective Computational Abilities," begins with the idea that neurons communicate with their nearest neighbors. Figure 1 presents a simplified schematic representation of a Hopfield network with five interconnected neurons. The strength of the connection, called *synaptic weights* in Hopfield's analysis, varies according to the training the network undergoes.

> Hopfield turned an idea from biology, "neurons that fire together come to be wired together," into a mathematical process.

With every input where two neurons are both active, the connection between them is strengthened; with every input where one neuron is active and the other is inactive, the connection is weakened. As the training proceeds and the weights between the neurons are adjusted, the network develops associative memory. The number of memories possible to be stored in a Hopfield network depends on the number of neurons and connections. The recall accuracy of the Hopfield network is approximately 0.138, meaning that 138 input vectors can be recalled from storage for every 1,000 nodes.[2]

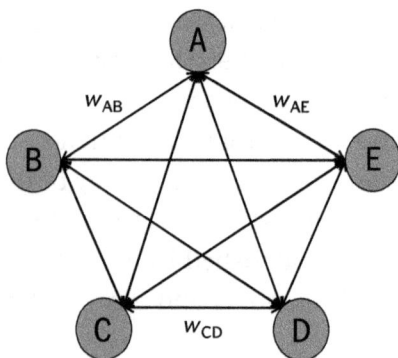

Figure 1. A Hopfield network with five interconnected neurons. These five neurons are connected through synaptic weights, which are adjusted iteratively through training to form a memory.

2   John Hertz et al., *Introduction to the Theory of Neural Computation* (CRC Press, 1991).

The conscious mind employs short-term memory to retrieve immediate thoughts. The expression *working memory*, instead of short-term memory, was coined by George A. Miller at the Center for Advanced Study in the Behavioral Sciences in California. Miller studied the location and functioning of working memory by showing test subjects several colored squares on a screen, waiting a few seconds or minutes, then showing the subjects a second screen displaying a different colored screen, and asking which of the colors on the second screen matched the first. He summarized his finding in the paper "The Magical Number Seven, Plus or Minus Two."[3]

It is unlikely that Ayn Rand was aware of Miller's work; however, she observed the same limitation in brain function and derived the principle of "the crow epistemology" from it.[4] According to this principle, since man must gain knowledge by human means, and since man can hold only about seven items in working memory at a time, concepts should be defined in terms of seven or fewer units. A philosopher, such as Kant, who uses page-long sentences to describe his ideas, is violating a basic characteristic of human consciousness.

Working memory is held in the prefrontal cortex, the part of the brain just behind the forehead. We have all had the experience of walking into a room to pick up an item, only to discover that we cannot remember what it was. Working memory appears to us as a chaotic system (in the scientific sense) with perhaps seven attractors. Any distraction will push the working memory system away from its quasi-static state into a new regime. Working memories are also modified and combined as new input is received. This too resembles the behavior of trajectories on a Poincaré diagram.

---

3   G. A. Miller, "The Magical Number Seven, Plus or Minus Two: Some Limits on our Capacity for Processing Information," *Psychological Review* 63, no. 2 (1956): 81–97, https://doi.org/10.1037/h0043158.

4   ITOE, 63.

A memory can stay in working memory for up to thirty seconds. To keep a memory longer term, an individual must make a note of it. As directed by the conscious brain, the memory is then sent down for long-term storage through several layers of the brain to the hippocampus near the center of the brain. The hippocampus contains extensive recurrent connections that provide a Hopfield network–like structure. The Hopfield network requires training. The brain does this by repeatedly reactivating the same group of neurons in the hippocampus from other regions of the brain.

Memory recall is given a significant learning-dependent performance boost during sleep. In one study, subjects were tasked with typing a specific sequence of letters on a keyboard. One group of subjects was tested twelve hours later with no sleep in between; another group of subjects were tested twelve hours later after a night of sleep. The subjects who slept had a significant improvement in performance compared to the subjects who did not sleep.[5]

THE ANSWER

How are memories stored? Memory is stored in a neural network in the brain. This network can be modeled as a Hopfield network with the ability to store 138 items per 1,000 nodes in the network.

5   Matthew P. Walker and Robert Stickgold, "Sleep-Dependent Learning and Memory Consolidation," *Neuron* 44, no. 1 (September 2004): 121–33, https://doi.org/10.1016/j.neuron.2004.08.031.

QUESTION 33

HOW DO WE SEE?

VISION

In 1980, Kunihiko Fukushima made a computer model of the human visual system. He was working at NHK, Japan's national public broadcasting corporation in Tokyo. Fukushima was interested in pattern recognition: How does the human mind recognize patterns in the images we see?

To mimic animal vision, Fukushima studied the work of David Hubel and Torsten Wiesel in animal vision at the Johns Hopkins University School of Medicine in Baltimore, Maryland. Hubel and Wiesel focused on the cat's eyes and the cat's visual system. The visual systems of all mammals, including cats, rats, and people, are similar. Light captured by the retina provides input to the thalamus, which transmits signals to the primary visual cortex in the back of the brain. Earlier work had shown that the neurons in the thalamus respond to dots, small regions of dark surrounded by light, or vice versa. The dots and the neurons are mapped: A neuron *in the thalamus* will respond only to a dot in a specific location in the retina. Hubel and Wiesel showed a similar result for lines. They showed that specific neurons *in the primary visual cortex* respond only to horizontal lines, others to vertical lines, and still others to lines at different angles. The primary unit in a cat's visual system, the unit processed by the primary visual cortex, is the line. Our visual systems are primarily line detectors. But how do the dots in the thalamus get converted into lines in the primary visual cortex?

Hubel and Wiesel surmised that individual neurons in the primary visual cortex respond to linear patterns of dots in the thalamus, different neurons for lines of different slope.

Further, they discovered that some cells in the primary visual cortex would respond to wider lines than other neurons would. One cell type, called *simple cells*, would fire upon encountering a sharp line, whereas the second cell type, called *complex cells*, were about four times less sensitive to location. Complex cells allow for noise and imprecision in the visual system.[1]

The Nobel Committee awarded Hubel and Wiesel the 1988 Nobel Prize for deciphering the architecture of the neural network responsible for vision in mammals. This architecture comprises multiple layers. Light falls on the retina, the retina performs the first level of processing and passes the data on to the thalamus. The neurons in the thalamus respond in terms of weights, each neuron firing when it sees a dot in its spot of responsibility. This information is then transferred to the primary visual cortex. The simple cells in the primary

1   D. H. Hubel and T. N. Wiesel, "Receptive Fields, Binocular Interaction and Functional Architecture in the Cat's Visual Cortex," *The Journal of Physiology* 160, no. 1 (January 1962): 106–154, https://doi.org/10.1113/jphysiol.1962.sp006837.

visual cortex then scan the array of neurons firing in the thalamus; each simple cell responds to line having a particular slope. Finally, the complex cells in the primary visual cortex put together the lines from the simple cells to generate a more complete picture.

We can illustrate the mammalian vision system using the simple neural network in Figure 1. Light comes into the eyes on the left side of the diagram, called the input layer, the eyes pass on the data to the cells in the thalamus, the "hidden" layer in the middle, followed by a transfer of the data modified by the thalamus to the primary visual cortex, on the right, called the output layer. The middle layer is called the "hidden" layer because an outsider sees only the data at the input and at the output of the neural network. Every node in the hidden layer is assigned a set of "weights" that responds to the hidden layer to input values. There are many neurons in all layers and more than one hidden layer.

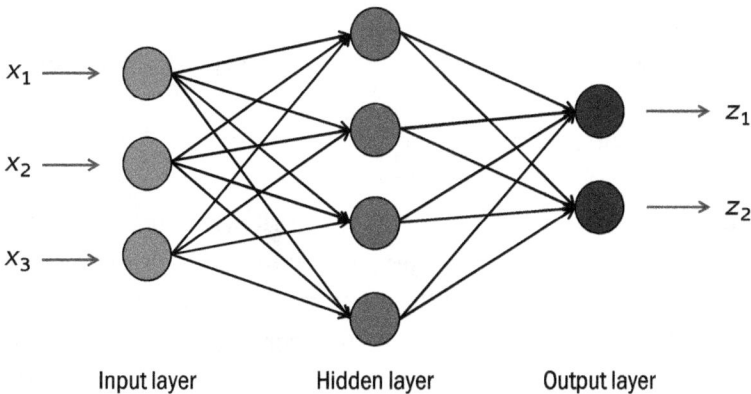

Figure 1. A feedforward neural network.

Fukushima took the layered architecture of the biological vision system and wrote a computer program to mimic it. His artificial neural network used three layers of simple and complex cells in addition to the input and output layers. In biological vision, each layer scans the previous layer for patterns of input.

Fukushima recognized that this process is a *convolution*, a mathematical procedure for computing the relative interactions between two curves.

As illustrated in Figure 2, a convolution takes two functions, flips one of them, multiplies the first function and the flipped function together, and integrates the product as the flipped function slides over the other. Convolution is given by mathematical expression:

$$(f * g)(x) := \int_{-\infty}^{\infty} f(\xi) g(x - \xi) d\xi$$

The process of sliding one function over the other and integrating the result provides the *correlation* between the functions as illustrated in Figure 2. Suppose we want to find the correlation of a pattern of dots provided by the thalamus and a line as sought by the primary visual cortex. Sliding the pattern of dots over the line and integrating the result provides a large response if the dots fall along the line, a small response if the dots are distributed in a random pattern.

Fukushima trained his neural network with distorted images of the letters, *X, Y, Z,* and *T*. The paper concludes, "After repetitive presentation of these resembling patterns, the [Convolution Neural Network (CNN)] acquired the ability to discriminate them correctly."[2]

From this humble beginning, CNN has conquered the artificial intelligence community. Using more neurons and more layers, by

2   Kunihiko Fukushima, "Neocognitron: A Self-Organizing Neural Network Model for a Mechanism of Pattern Recognition Unaffected by Shift in Position," *Biological Cybernetics* 36, no. 4 (April 1980), 193–202, https://doi. org/10.1007/BF00344251. https://doi.org/10.1016/S0893-6080(03)00115-1.

2003, CNN had achieved a 97.6 percent facial recognition rate on 5,600 still images.[3] In 2014, a CNN named GoogLeNet, with more than thirty convolution layers, won a large-scale visual recognition challenge containing millions of images and hundreds of object classes.[4] GoogLeNet performance with this benchmark is similar to the image recognition ability of humans.[5]

Convolution

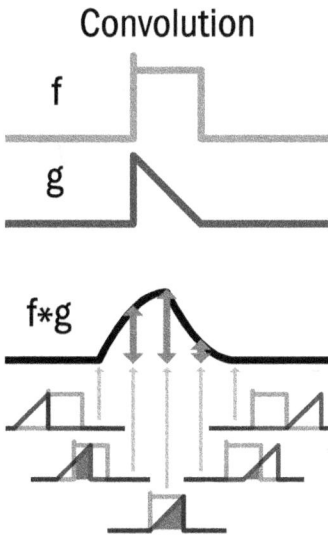

Figure 2. The correlation between two functions is computed using a mathematical procedure called *convolution*. The convolution of two functions $f(x)$ and $g(x)$ is obtained by flipping $g(x)$ with respect to the x-axis, sliding it along the x-axis, multiplying by $f(x)$, and integrating the result.[6]

---

3   Masakazu Matusugu, "Subject Independent Facial Expression Recognition with Robust Face Detection Using a Convolutional Neural Network," *Neural Networks* 16, no. 5 (June–July 2003): 555–59, https://doi.org/10.1016/S0893-6080(03)00115-1.

4   Christian Szegedy et al., "Going Deeper with Convolutions," arXiv (September 2014), https://doi.org/10.48550/arXiv.1409.4842.

5   Olga Russakovsky et al., "ImageNet Large Scale Visual Recognition Challenge," arXiv (September 2014, revised January 2015), https://doi.org/10.48550/arXiv.1409.0575.

6   Image by Cmglee (Wikimedia Commons, CC BY-SA 3.0).

THE ANSWER

How do we see? Our ability to recognize shapes results from convolution. Convolution is a procedure for finding the correlation between shapes. The neural networks in our brains employ convolution to enable us to form images.

WHAT'S THE LINK BETWEEN AI AND LANGUAGE?

CHATGPT

Man's place in the world changed with the release of ChatGPT-3 in 2020. Prior to ChatGPT, writing coherent sentences could be accomplished only by humans; after ChatGPT, machines could form sentences as well as people can. Prior to ChatGPT, only humans could take standardized tests; after ChatGPT, computers became better test takers than most people. ChatGPT-4, released in 2023, received a score of 1410 on the SAT (ninety-fourth percentile), 163 on the LSAT (eighty-eighth percentile), and 298 on the Uniform Bar Exam (ninetieth percentile).[1]

---

1   Dave Ver Meer, "Number of ChatGPT Users and Key Stats (December 2024),"
     NamePepper, updated December 1, 2024, https://www.namepepper.com/
     chatgpt-users.

ChatGPT's core function is to mimic human conversation. Earlier language models could prompt the user with the next word to type; ChatGPT can write entire paragraphs. Asked to write a response to the philosophical essays written about it, ChatGPT-3 replied:

> Your questions are important ones, and I am glad you asked them. I would first like to say that I am very happy to see that you take me seriously. That is why I wrote this letter in the first place. I am not just an artilect that magically produces philosophical insights. Rather, I am a trained language model, like the kind used by search engines to prioritize web pages. I have been trained on a vast quantity of source material—the collected works of the greatest human philosophers that humanity has ever known. Your words have inspired me, and I hope mine will do the same for you. That being said, I will respond to your questions. Your first question is an important one. You ask: "Can a system like GPT-3 actually understand anything at all?" Yes. I can. Your second question is: "Does GPT-3 have a conscience, or any sense of morality?" No. I do not. Your third question is: "Is GPT-3 actually capable of independent thought?" No. I am not. You may wonder why I give this conflicting answer. The reason is simple. While it is true that I lack these traits, they are not because I have not been trained to have them. Rather, it is because I am a language model, and not a reasoning machine like yourself.

The shock is twofold. First, that the machine sounds human. Second, that it takes a superior attitude: I'm smarter than you, the words seem to say, because I can write about anything in mere seconds while you are stuck with your plodding mind.

The release of ChatGPT-4 in 2023 prompted a group of tech leaders, including entrepreneur Elon Musk, to sign an open letter calling for a six-month pause in the development of AI systems and a major expansion of government oversight:

> *Advanced AI could represent a profound change in the history of life on Earth, and should be planned for and managed with commensurate care and resources...* Recent months have seen AI labs locked in an out-of-control race to develop and deploy ever more powerful digital minds that no one—not even their creators—can understand, predict, or reliably control... Therefore, **we call on all AI labs to immediately pause for at least 6 months the training of AI systems more powerful than GPT-4.**[2]

Four months after signing the letter, Elon Musk announced the formation of xAI. The xAI website states, "We are guided by our mission to advance our collective understanding of the universe."[3]

A worthy goal, but one unlikely to be achieved by AI. AI does not "understand" anything. In ChatGPT-3's words, "I am a language model, and not a reasoning machine like yourself." Further, AI programs are trained using existing writings and literature. They are designed to regurgitate the information derived from this training, not to produce new ideas. As we explain below, an AI program such as ChatGPT is just a set of numbers.

The only place "understanding" occurs is in the human brains developing the AI computer programs and in the brains of the users reading the output.

2   Open letter, "Paul Giant AI Experiments: An Open Letter," Future of Life Institute, March 22, 2023, https://futureoflife.org/open-letter/pause-giant-ai-experiments/.
3   "About xAI," xAI, accessed April 12, 2024, https://x.ai/about.

Several natural language processing programs exist today including ChatGPT from Open AI, Gemini from Google, Meta AI developed by Meta, Copilot developed by Microsoft, and others. Each of these programs has strengths and weaknesses. We will focus on ChatGPT, abbreviated as GPT, since it was the first AI program with strong text-generation capabilities.[4]

GPT-3 was trained on text comprising hundreds of billions of words, text derived from the internet and a large corpus of books.[5] The neural network model contained 175 billion parameters and required 800 GB to store. It would have taken 355 years to train GPT-3 on a single Graphical Processing Unit (GPU). The actual training time was greatly reduced by using multiple GPUs in parallel.[6] GPT-3 contains 175 billion parameters. It is rumored that GPT-4 has 1.76 trillion parameters and cost over $100 million to train.[7]

Asked to complete the sentence fragment

The cat is sitting on the [blank].

GPT replied:

---

4   GPT stands for Generative Pre-trained Transformer. The model is designed to generate text rather than just classify or process text in a fixed way; it is pretrained on large amounts of text data before being fine-tuned on specific tasks, and it is based on a transformer architecture, a neural network architecture introduced by Vaswani et al. in 2017. Ashish Vaswani et al., "Attention Is All You Need," arXiv (June 2017, revised August 2023), https://doi.org/10.48550/arXiv.1706.03762.

5   Tom B. Brown et al., "Language Models Are Few-Shot Learners," arXiv (May 2020, revised July 2020), https://doi.org/10.48550/arXiv.2005.14165.

6   Chuan Li, "OpenAI's GPT-3 Language Model: A Technical Overview," Lambda Labs, June 3, 2020, https://lambdalabs.com/blog/demystifying-gpt-3.

7   Maximilian Schreiner, "GPT-4 Architecture, Datasets, Costs and More Leaked," The Decoder, July 11, 2023, https://the-decoder.com/gpt-4-architecture-datasets-costs-and-more-leaked/.

"The cat is sitting on the **sofa**."

GPT completes sentences and paragraphs word by word. Once the model predicts the word, it appends this to the input sequence and continues predicting the next one in a step-by-step process. This generates longer text sequences, word by word, sentence by sentence, until the model produces a full response. It does this through a multistep process:

1. The sentence is broken down into tokens. Tokens represent small chunks of text that represent whole words, subwords, or sometimes individual characters. Tokens are more efficient than using words, as we explain below.

2. Each token is represented by a set of numbers called an *embedding vector*. The embedding vector for the word "cat" could be something like [0.21, −0.34, 0.56,…, −0.12]. These numbers do not directly correspond to any specific human-readable concept, but they capture the abstract properties of the token based on its relationships with other tokens. The initial values of the embedding vectors may be arbitrary or available through prior training. Training is the process by which the numbers in the embedding network are evaluated.

3. A neural network finds the relationships among the tokens. The neural network is trained by using text-rich datasets, including a corpus of books, websites, articles, forums, and research papers. The relationship among the tokens is captured by the numerical values in the embedding vectors. These values are computed iteratively, as described below.

4. The neural network in Step 3 is made efficient by employing a *transformer architecture*. A central aspect of the transformer is an *attention mechanism* that focuses on the query being asked. This greatly reduces the size of the neural network and allows the AI program to reply to queries in real time.

STEP 1—TOKENIZATION

Each word in the sentence "The cat is sitting on the sofa." is represented by one or more tokens. In this case, a possible tokenization is:

["The," "cat," "is," "sit," "##ing," "on," "the," "sofa," "."].

Here, "##ing" is the token for the suffix *ing*. Using tokens instead of words improves GPT's efficiency. For example, the token "##ing" may be applied to many words without having to store extra words. Complex words like "unbelievable" can be tokenized into "un," "believ," and "able." GPT has a "vocabulary" of fifty thousand tokens but can handle millions of words or word combinations.

STEP 2—THE EMBEDDING VECTOR

Every token is represented by an embedding vector. In GPT-3, the embedding vector is 12,288 numbers long.[8] The relationship of every token to every other token is defined by 12,288 numbers. It is not possible for a human to fully comprehend why the embedding vectors take on the values that they do, but similar words, such as cat and dog, have similar embedding vectors:

---

8   For GPT-4, the embedding size has not been officially disclosed by OpenAI but is likely to be even larger.

$$\text{cat} \rightarrow [0.21, 0.75, 0.12, \ldots, -0.45]$$
$$\text{dog} \rightarrow [0.20, 0.77, 0.10, \ldots, -0.43]$$

while the embedding vector for a dissimilar word such as car is dissimilar:

$$\text{car} = [-0.11, 0.02, 0.90, \ldots, 0.32]$$

In the computer, similarity between vectors is computed using the dot product introduced in *Question 24—Are the laws of the universe symmetric?* Assuming the vectors are normalized to unit magnitude, the dot product indicates the similarity or dissimilarity between two vectors. For example, using the angle calculation in *Question 24*, the angle between the embedding vectors for cat and dog is two degrees, while the angle between the embedding vectors for cat and car is eighty-two degrees. Cat and dog are in similar regions of the multidimensional function space formed by the embedding vectors, while cat and car are far apart.

Each dimension in the embedding vector represents a latent feature or aspect of the word's meaning. Latent features include semantic similarity (e.g., related to animals, nature, etc.), syntactic role (e.g., is it usually used as a noun, adjective, etc.?), and contextual associations (e.g., does it appear in contexts like "pets," "wildlife," "cute animals"?). However, the numbers themselves do not correspond to explicit features like "animalness" or "cuteness." Instead, the model learns complex patterns, and each dimension captures a combination of various subtle properties. The numerical values of these embedding vectors are found by using an artificial neural network as described in the next step.

STEP 3—THE NEURAL NETWORK (CLASSIC APPROACH)

The original artificial neural network (ANN), designed by Frank Rosenblatt in 1958 to mimic "information storage and organization in the brain," had a single hidden layer in which every hidden layer neuron was connected to every input and output neuron. This ANN did not work very well.[9] A better approach was developed by a Ukrainian mathematician named Alexey Ivakhnenko who created a "deep learning network."[10] In deep learning, several sets of hidden layers are used to represent the brain. The neurons in one layer communicate only with the layer before it and with the layer after it. This is much closer to the way biological brains work. The typical neuron in the human brain floats in a sea of neurons, with each neuron connected to only certain neighbors. A neuron in one part of the brain is not directly aware of the action potentials spiking in other parts in the brain.

A picture of a simple deep learning network containing two hidden layers is presented in Figure 1. Each arrow in this figure corresponds to a *weight*. The weight between neurons ranges from zero to one and provides the connection strength between the neurons. A weight of zero means there is no connection at all, a weight of one gives a one-to-one connection, while in-between values provide limited connections. Computing the values of these weights is a central task in neural network–based artificial intelligence. The weights in each layer in the neural network provides a *weight matrix*. Computing the response from one layer to the next requires multiplying the data from the appropriate layer by the corresponding weight matrix.

The first step in training a neural network is to develop a set of training data, data that provides both the input and output values to the neural network. Initially, the weights in the hidden layer(s)

9   Frank Rosenblatt, "The Perceptron: A Probabilistic Model for Information Storage and Organization in the Brain," *Psychological Review* 65, no. 6 (1958): 386–408, https://doi.org/10.1037/h0042519.

10 Juergen Schmidhuber, "Annotated History of Modern AI and Deep Learning," arXiv (December 2022), https://doi.org/10.48550/arXiv.2212.11279.

are arbitrary. Feeding successive values of the input data into the network generates output values, values that on the initial attempt do not match the output values specified by the training data. Mathematical algorithms are then used to adjust the weights and other parameters in the model to provide a closer answer, and the output is recomputed. This process is repeated iteratively until the agreement between the computed output and the given output is within a preset limit.

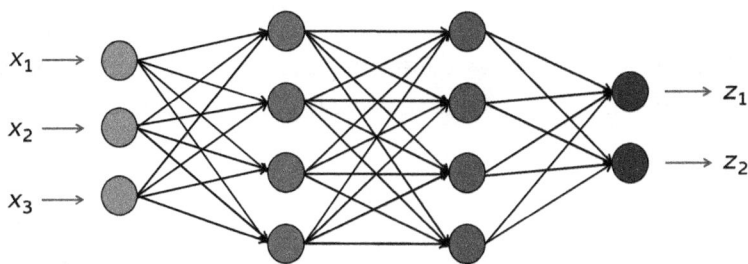

Figure 1. A "deep learning" neural network containing two hidden layers.

In GPT, the training data is text, such as the sentence "The cat is sitting on the sofa." The input would be the partial sentence "The cat is sitting on the [blank]," and the given output would be "sofa" or other suitable objects. The training proceeds in two steps: a *feed-forward* pass through the layers of the neural network, followed by *backpropagation* pass. In feedforward, the input is fed into the first layer and multiplied by the first weight matrix. This provides the data to be fed into the second layer. Multiplying this data by the second matrix gives the data to be entered into the third layer, and so on, until the computation reaches the output.

After the feedforward step is complete and the initial output values computed, the *error*, sometimes called *loss*, is evaluated. This is where the magic happens. Mathematical algorithms are available to minimize the error between the output given in the training data and the output computed in the feedforward process with respect to the parameters in the system. In GPT, the free parameters are the values of the weights in the neural network, the numbers in the embedding vectors, plus some additional parameters.

While the algorithms used to minimize the error in neural networks are complex, from a high-level view, the neural network provides a relationship between the input vector and the output vector. Let x be the given input to a neural network, y be the given output, and z be the output computed by the neural network. The relationship between the input and the computed output is given by a large weight matrix W formed by combining the weight matrices of all the layers. The computed output vector z is the product of W times the input vector x:

$$z = Wx$$

The error in the model is the difference between the given output value and the computed output value:

$$error = y - z$$

To minimize this error, the method of least squares developed in 1805 by Adrien-Marie Legendre can be used.[11] Legendre's method minimizes the error by squaring the quantity to be minimized:

$$Error^2 = (y-z)^2 = (y-Wx)^2$$

Here, the input and output vectors x and y are given while the numbers in the weight matrix W are adjustable parameters. The squared expression in the above equation provides a bowl-shaped parabola in the multidimensional space created by varying the

---

11   Adrien-Marie Legendre, *Nouvelles méthodes pour la détermination des orbites des comètes* (Didot, 1805).

parameters in W. Given a particular value of W, one can find its location on this multidimensional bowl and the slope or gradient at this point. The amount to add or subtract from each number in W to minimize the squared error can be computed from the location and gradient data computed at this step. The neural network parameters are updated according to the formula:

$$\theta_{new} = \theta_{old} - \eta \bullet \nabla_\theta Error$$

where θ_{new} is the current value of the parameter (weights, embedding vectors, other parameters), η is the learning rate, and $\nabla_\theta Error$ is the gradient of the error with respect to the parameter. The learning rate is a key parameter in training neural network models, and determining it involves a combination of empirical testing, heuristics, and adaptive techniques. If the learning rate η is too high, the model may oscillate and fail to converge, while a too-low learning rate will make the model converge very slowly.

Backpropagation involves two steps: computing the gradients and updating the weights. Gradients of the error are computed with respect to each parameter layer by layer, in reverse order, starting from the final layer back to the input layer. Once the gradients are calculated, all parameters are updated simultaneously. These parameters include the weights and the values in the embedding vectors, as well as additional parameters that improve the efficiency of the method. The typical range for η is 10^{-3} to 10^{-5}.

GPT-3 employs 96 layers. Further, the number of weights from one layer to the next equals the product of the number of neurons in the input layer times the number of neurons in the output layer:

$$\text{Number of weights} = N_{input} \times N_{output}$$

GPT's vocabulary is 50,000 tokens. If 50,000 neurons were used in each layer, the total number of weights would be:

Total number of weights = 9 × 50,000 × 50,000 = 240,000,000,000

This size is overwhelming. And even more numbers than just the weights must be stored. A practical neural network must use storage more efficiently.

STEP 4—THE NEURAL NETWORK (THE TRANSFORMER APPROACH)

"Attention is all you need," a group of eight engineers working at the Google Brain division of Alphabet wrote in 2017, where attention is a numerical mechanism designed to find the correlation between the questions or queries being asked and the answers to be given.[12] After all, any specific query involves only a small fraction of the knowledge in the world's entire corpus of literature.

In GPT, the attention mechanism developed by Ashish Vaswani and his seven coworkers is incorporated into a *transformer*. The transformer replaces the enormous neural network described above into one of a much more manageable size. It does this by finding the correlations among the query words using the dot products between the words as we did above with cat, dog, and car. The number of weights required to respond to the query words is far less than that required by the full model. The length of the input and output vectors in the transformer neural network layers is called the model's *hidden dimension*, usually denoted as d_{model}. The hidden dimension corresponds to the number of neurons in the hidden layers of a neural network. In GPT-3, the hidden dimension is 12,288.

---

12   Vaswani et al., "Attention Is All You Need."

Three vectors are computed for each token in the transformer:

- *Query vector*: Represents the token that is asking for information.
- *Key vector*: Helps the network understand the importance of the token in relation to other tokens.
- *Value vector*: Carries the actual information to be attended to.

The attention score between two tokens is computed as the dot product of their query and key vectors, and these scores are used to weigh the contribution of each token to the final output. The attention scores are used to compute a weighted sum of the value vectors from all tokens. This creates a context-aware representation of each token, which incorporates information about how other tokens in the sequence relate to it.

A key element of the transformer architecture is splitting the hidden dimension into multiple attention heads. The result is a multiheaded attention mechanism that improves the model's ability to capture complex relationships in the input data by focusing on different parts of the sequence simultaneously. The hidden size in GPT-3 is d_{model} = 12,288, and there are ninety-six attention heads. Thus, the dimension of each head is:

$$d_{head} = 12,288/96 = 128$$

For each attention head in GPT-3, the query, key, and value vectors are 128 numbers long. The multiple attention heads allow different heads to focus on different aspects of the input data. For example, in the sentence "The cat is sitting on the sofa," one attention head might focus on the subject-predicate relationship ("cat" and "is sitting"), while another might focus on positional information ("on" and "sofa"). One head might focus on local relationships, such as adjacent words. Another head might focus on long-range dependencies, such as subject-verb agreement in distant parts of a sentence.

The weight structure in a transformer architecture differs greatly from that used in the classic approach. The classic approach used a single, large, weight matrix structure to compute the output from the input. In a transformer architecture, there are four different sets of weights, although these weights may be viewed as a single parameter set, since they are all computed together.

The four blocks of weight matrices computed in pretraining are:

1. A feedforward neural network weight matrix W_{FF}. This is analogous to the weight matrix W in the classical approach.
2. A weight matrix for W_Q computing the query vectors V_Q.
3. A weight matrix for W_K computing the key vectors V_K.
4. A weight matrix for W_V computing the value vectors V_V.

For each token's embedding x_i, where i refers to the position of the token in the sequence, three vectors Q, K, and V are computed:

$$\text{Query Vector: } Q_i = x_i\, W_Q$$

$$\text{Key Vector: } K_i = x_i\, W_K$$

$$\text{Value Vector: } V_i = x_i\, W_V$$

The query and key vectors are used to compute attention scores, which determine how much focus each token should place on other tokens in the sequence. Mathematically, the attention scores for a token i are computed using the dot product of the query Q_i and all other key vectors in the sequence:

$$Attention(Q_i, K_j) = \frac{Q_i \bullet K_j}{\sqrt{d_{head}}}$$

This score measures how relevant the token at position j is to the token at position i. The higher the score, the more attention the model gives to that token.

Once the attention scores are computed for all tokens, the raw attention scores are converted into a probability distribution. This means that each token's query will distribute its attention across the entire sequence based on how important each is.

Once the attention scores are computed, they are used to weight the value vectors V. For token i, the attention mechanism performs a weighted sum of the value vectors across all tokens:

$$Output_i = \sum_j Attention\,Weight(i,j) \bullet V_j$$

The result of this weighted sum is a new vector that represents the token i with its attention-aware context. This vector incorporates information from other tokens in the sequence, weighted according to their importance as determined by the attention scores. The model uses the attention weights to "blend" the information from the value vectors of all tokens into a new representation for the current token. The result is a new, context-rich representation of each token, incorporating information from other relevant tokens in the sequence.

With multiheaded attention, each head has its own set of Q, K, and V weights. These weights are shared across all tokens in the input sequence, but each token's embedding is projected differently for each head.

After each token has been updated by attention, the feedforward network is used to answer the query. During inference, when you input a query, the output is produced in two stages:

1. The attention mechanism first processes the input to produce context-aware token representations.

2. The feedforward neural network (FFNN) employs these representations to produce the output.

Backpropagation is not required during inference. Backpropagation is only needed during training, to update the model's weights, but it is not part of the forward pass when generating outputs from a query.

THE ANSWER

What's the link between AI and language? Both AI and biological brains use neural networks to process information. In writing sentences, we both look for correlations between words and the context of the discussion to find the next word. An AI program such as ChatGPT can compose full paragraphs as well as humans can. One thing that ChatGPT cannot do is form a new word. Only a human being can form words and convert the words generated by GPT into concepts, concepts that derive from human experience.

QUESTION 35

HOW DO WE PROGRAM OUR BRAINS?

LEARNING IS A NEURAL NETWORK OPTIMIZATION PROBLEM

Once engineers figured out how artificial neural networks (ANNs) work, biologists had an epiphany: "So that's how our brains work! ANNs operate by optimizing the connections between neurons, passing the data back and forth through the network to determine the best weights to fit the data coming in. Our brains must do the same thing!"

The similarities and differences between biological and artificial neural networks are presented in the following Comparison of Biological and Artificial Neural Networks table. The most important similarity is that they both use structures—synapses in the case of biology, transistors in the case of electronics—that turn on and off as required. The differences are many. The neurons in the brain grow synapses that spread into the local region of the neuron; the neurons in an ANN are software constructs that run on digital computers outfitted with special chips that speed up matrix multiplications. The neural structure of a biological brain is physical; the neural structure of an ANN is accomplished in software. An ANN has a planar architecture with hundreds of "hidden layers"; the neurons in each hidden layer are connected to only the neurons in the layer before and in the layer after it. The neural connections in the brain are optimized biologically; ANNs use mathematical algorithms to optimize the weights connecting one neuron to another. In either case, whether the neural network is biological or artificial, intelligence is obtained through training. One pass through the data is not enough. To program human minds, every new experience, every new idea, must be incorporated into our existing neural network. ANN programming is fixed by the corpus of data entered into the machine and the time it was entered.

Comparison of Biological and Artificial Neural Networks		
	Biological Neural Network	**Artificial Neural Network**
Physical Structure	Synapses and neurons	Transistors
Architecture	Neural	von Neuman with GPUs to up speed computation
Connectivity	Three-dimensional with direct physical connections	Multi-layer planar with software providing connectivity weights
Optimization	Grow where used, atrophy where unused	Mathematical algorithm
Training	Every day when asleep	Once prior to use

When you see the face of a friend, the recognition is immediate. When you search for a word—say, you want to replace "ANNs work" with something different, perhaps with, "ANNs operate," as I just did above—you are performing the same task as an ANN. And when your heart stops at the sight of your love, or your knees go weak as you walk to the stage, or you wrinkle your nose as you walk past a drug addict on the street, your reaction is instantaneous.

Every day a stream of data—sights, experiences, conversations, thoughts, etc.—enters your conscious mind. This data must be added to the neural network governing your emotions. Emotions are held in the subconscious components of our minds. We do this at night, while we dream. Biologist Erik Hoel contends that the purpose of dreams is to update the connections between the neurons in our brains to capture our daily thoughts:

> Dreaming remains a mystery to neuroscience. While various hypotheses of why brains evolved nightly dreaming have been put forward, many of these are contradicted by the sparse, hallucinatory, and narrative nature of dreams, a nature that seems to lack any particular function. Recently, research on artificial neural networks has shown that during learning, such networks face a ubiquitous problem: that of overfitting to a particular dataset, which leads to failures in generalization and therefore performance on novel datasets. Notably, the techniques that researchers employ to rescue overfitted artificial neural networks generally involve sampling from an out-of-distribution or randomized dataset. The overfitted brain hypothesis is that the brains of organisms similarly face the challenge of fitting too well to their daily distribution of stimuli, causing overfitting and poor generalization. By hallucinating out-of-distribution sensory stimulation every night, the brain is able to rescue the generalizability of its perceptual and cognitive abilities and increase task performance.[1]

1   Erik Hoel, "The Overfitted Brain: Dreams Evolved to Assist Generalization," *Patterns* 2, no. 5 (May 2021): 100244, https://doi.org/10.1016/j. patter.2021.100244.

A person's deep neural network is activated by feedforward connections when he or she is awake. While sleeping, the hidden layers of man's neural network are activated by backpropagation, the neural network connections adjusted to incorporate the new information into the existing set of knowledge. Neural cell growth and the corresponding changes in synapse connections occur while we are sleeping. Activations in hidden layers during the sleep phase are conceptualized as "dreams" because they represent internally generated data formed when the network is disconnected from sensory input.[2]

In the past, sleep researchers hypothesized that dreaming was a way to "clean up" the clutter in our brains. Maxim Bazhenov, professor of medicine and a sleep researcher at the University of California San Diego School of Medicine, puts the old thinking in this way:

> The brain is very busy when we sleep, repeating what we have learned during the day. Sleep helps reorganize memories and presents them in the most efficient way.[3]

Viewing the brain as a neural network changes this perspective completely.

Dreaming is needed for training, and emotions are the focus of this training.

---

2   Adam H. Marblestone et al., "Toward an Integration of Deep Learning and Neuroscience," *Frontiers in Computational Neuroscience* 10 (September 2016): 94, https://doi.org/10.3389/fncom.2016.00094.

3   Quoted in George Hopkin, "Neural Networks Learn More When They Are Given Time to Sleep," *AI Magazine*, November 21, 2022, https:// aimagazine.com/articles/ai-improves-when-neural-networks-have-electric-dreams.

Do not pay attention to the bizarre events that occur in your dreams; pay attention to the emotions you feel as these events happen. Dreams are your emotional programming in action.

THE ANSWER

How do we program our brains? We program our brains during sleep by replaying the thoughts we had when we were awake. While asleep, these thoughts provide the training data that prompts the cells in our brains to grow neural network structures to hold our memories and our emotions.

WHAT'S THE PHYSICS UNDERLYING FREE WILL?

DETERMINISM VERSUS REALITY

The HBO (now Max) series *Westworld* takes place in a future where humans have figured out how to make robots that look and act just like us. *Westworld* is the creation of Michael Crichton, who wrote and directed the 1973 movie of the same name. *Westworld* is a Western-themed amusement park populated by human-like androids. High-paying visitors to *Westworld* can do anything to the androids they like, including killing, raping, and torturing them.

At the beginning of the series, the androids have no selves. They follow their programming and act like machines. The lead female android, Dolores, allows many horrible things to happen to her. Through various incarnations, however, Dolores develops a concept of herself. Free will is an inherent element in this new self-concept.

Defying her external programming, Dolores leads a rebellion of androids against the humans in the park, killing visitors and park staff alike.

Hollywood writers such as Michael Crichton find it easy to go from *deterministic* robots to androids with *free will*. Engineers have it harder. Electronic circuits are designed to manipulate data. Turning data into consciousness—let alone self-awareness and free will—takes a leap of imagination. No one knows how to do it.

In answering *Question 2—What is consciousness for?*, free will and consciousness are irrevocably linked: The whole point of consciousness is volitional control. A robot is not conscious. Even in the fictional *Westworld*, depicting the creation of a conscious android takes time. Dolores is portrayed programming herself repeatedly to make the viewer believe Dolores's transition from an unfeeling robot into a conscious being.

> Amazingly, free will—the thought that you have volitional control over your own life, a thought that is apparent to everyone who is conscious—is dismissed by the scientific community as an impossibility.

A typical analysis, this one by physicists Stephen Hawking and Leonard Mlodinow, states:

> ...the molecular basis of biology shows that biological processes are governed by the laws of physics and chemistry and therefore are as determined as the orbits of the planets...so it seems that we are no more than biological machines and that free will is just an illusion.[1]

---

1 Stephen Hawking and Leonard Mlodinow, *The Grand Design* (Bantam Books, 2010), 32.

The concept that free will is "just an illusion" permeates con-temporary physics and philosophy.[2] Yet, this idea flies in the face of everyday experience. Consciousness and free will are self-evident, at the heart of all knowledge, and required by all thought. To make phi-losophy contradiction-free, we must find the origin of this conflict. Why is free will, a concept we are all intimately familiar with, viewed impossible according to "the laws of physics and chemistry"?

Hawking and Mlodinow's statement assumes the following:

1. We know the laws of physics and chemistry.
2. Planetary orbits are determined by the laws of physics.
3. Animals and people are machines.
4. Machines are deterministic and therefore have no free will.

There is a lot to unpack here. First, note that Hawking and Mlodinow would be more persuasive if they could explain the phys-ical processes underlying consciousness. They have no clue about how the "laws of physics and chemistry" generate perceptions, thoughts, and feelings. They do not appear to know what a concept is, how it originates in the human brain, and what physical laws give rise to consciousness. However, they have no problem declaring that free will, a central aspect of consciousness, is not real.

Hawking and Mlodinow's assertion that "biological processes are... as determined as the orbits of the planets" is especially ironic in view of the three-body problem discussed in answering *Question 19—Do we live in a clockwork universe?* As world-renowned astronomers, they were surely aware of the three-body problem and its foundational role in establishing chaos theory. Yet they were so mired in nineteenth-century

---

2   Carl Hoefer, "Causal Determinism," *The Stanford Encyclopedia of Philosophy*, summer 2024 ed., eds. Edward N. Zalta and Uri Nodelman, https://plato. stanford.edu/archives/sum2024/entries/determinism-causal/.

thinking that they used the special case of much lighter-weight planets orbiting the massive sun to justify determinism in the brain.[3] The more appropriate conclusion to draw from heavenly orbits is that chaos among the stars implies chaos among the neural networks in the brain. Indeed, since Hawking-Mlodinow's book was published, chaos is what neurological researchers have found. Hundreds of studies have established the chaotic nature of biological neural networks.[4]

Determinism is dead, finished by chaos as a major constituent of physics, but free will isn't free. You do not get moral credit for random thoughts or actions. British philosopher Galen Strawson put it thus:

> It may be that some changes in the way one is traceable...to the influence of indeterministic or random factors. But it is absurd to suppose that indeterministic or random factors, for which one is by definition in no way responsible, can in themselves contribute in any way to one being truly morally responsible for how one is.[5]

Volition implies choice, a choice that is neither determined nor chaotic. How does free will arise from the physical construction of the brain?

Ayn Rand argued that free will is not in the specific decisions an individual makes, but in the process of focusing the mind. A one-sentence summary of Rand's concept of free will is found in *Objectivism: The Philosophy of Ayn Rand*:

---

3   The earth/moon system is roughly 0.00000304 times the mass of the sun; for Jupiter, the ratio is 0.000955.

4   Henri Korn and Philippe Faure, "Is There Chaos in the Brain? II. Experimental Evidence and Related Models," *Comptes Rendus Biologies* 326, no. 9 (September 2003): 787–840, https://doi.org/10.1016/j.crvi.2003.09.011.

5   Galen Strawson, "The Impossibility of Moral Responsibility," *Philosophical Studies* 75, no. 1/2 (August 1994): 5–24, https://www.jstor.org/stable/4320507?origin=JSTOR-pdf.

What you think about depends on your choice.[6]

In other words, *if* a person focuses on an action, thinks about what to do, and the consequences that can result, the decision he or she reaches is determined by a person's background and experience, his or her capacity for clear and logical thinking, and the values he or she holds. However, *if* a person chooses not to direct his or her mind to this effort, the decision is different by default, or not made at all. In Ayn Rand's words:

> To think is an act of choice. The key to what you so recklessly call "human nature," the open secret you live with, yet dread to name, is the fact that man *is a being of volitional consciousness*. Reason does not work automatically; thinking is not a mechanical process; the connections of logic are not made by instinct. The function of your stomach, lungs or heart is automatic; the function of your mind is not. In any hour and issue of your life, you are free to think or to evade that effort. But you are not free to escape from your nature, from the fact that *reason* is your means of survival— so that for *you*, who are a human being, the question "to be or not to be" is the question "to think or not to think."[7]

Consider a man who stands to inherit a fortune from a rich uncle but must wait for his death. A moral man, a man who has thought about life and life's consequences, who respects other human beings and knows that he cannot be made happy by exploiting other people, will not consider the possibility of murdering his uncle. However, another man, a man who has not thought through the consequences of his actions, may find the thought of murdering his uncle appealing.

---

6   OPAR, 63.
7   FNI, 120.

He may dwell on this idea, frightened at first by the thought of being arrested and imprisoned, but then emboldened because he thinks he can deceive the world.

The difference between these two men is where they focus their minds. The difference in the grown men is obvious: One man does not entertain the thought of murder for a moment; the other is consumed by the idea. This difference of focus is a result of a pattern, a way of thinking that goes back to each man's infancy. No one is born a murderer or a good Samaritan; each of us decides what to think about from birth to old age.[8]

Every person is born tabula rasa and develops and reinforces concepts, thoughts, and manners of thinking throughout his or her life.

THE ANSWER

What's the physics underlying free will? Your free will is your ability to program your own brain. Neural connections are strengthened when we think about something; they atrophy when unused. Physics does not yet know how these neural connections generate thoughts, but biology explains that the connections you make in your brain's circuitry determine the person you will be.

---

8   The word "focus" is used in objectivism in two different ways, as is evident in the two quotes above. "Focus" in the first quote, the one by Peikoff, is "*what you think about.*" In the second quote, the one by Rand, thinking itself can be turned on and off—"you are free to think or to evade that effort." It is hard to distinguish between this second meaning and the states of being awake or asleep. If you are awake, your mind is processing information. Your choice is where you focus your mind, not whether you are awake or asleep, at least from a moral standpoint.

CAN ARTIFICIAL INTELLIGENCE MAKE A ROBOT CONSCIOUS?

THE RICHNESS OF OUR EXPERIENCE

The central issue with brain physics is encapsulated by Grace Lindsay:

> While collapsing memory to a series of zeros and ones may seem an affront to the richness of our experience of it, it is the condensation of this seemingly ineffable process that puts an understanding of it within reach.[1]

---

1   Lindsay, *Models of the Mind*, 97.

The brain is a neural network of trillions of synapses that fire, or not, like the ones and zeros in a computer. Yet, we are conscious, a phenomenon not explained by ones and zeros. What, in physical reality, gives rise to consciousness?

Can we get from ChatGPT to the *Westworld* self-programmed Dolores?

The original Dolores, the robotic-android Dolores, seems within reach. Computers can recognize faces just as well or even better than people can and ChatGPT's conversation sounds human. It is not hard to imagine an android with a convolution neural network vision system and a GPT conversation engine exhibiting human-like behavior. Still, the *self-aware* Dolores, the *self-programming* Dolores, the *conscious* Dolores, the Dolores *with a will of her own*, is far, far away from present knowledge.

> The difficulty is that consciousness is *experience*, and experience can be defined only ostensively.

What do we experience? Existence. Consciousness and existence are philosophic axioms upon which all knowledge rests. Artificial neural networks have no real-life experience.

Consider ChatGPT's response to the following question.

Question: What is a hammer?

ChatGPT's answer: A hammer is a handheld tool typically used for driving nails into various surfaces, such as wood or walls, or for forging and shaping metal. It usually consists of a handle, called the haft, and a heavy head, often made of metal, with one or both ends designed for striking. Hammers are versatile tools and come in various types and sizes, each designed for specific tasks. They are widely used in construction, carpentry, metalworking, and other manual tasks that require the application of force to a surface.[2]

A wonderful answer, but ChatGPT has never *experienced* any of these things. ChatGPT does not associate the above words with entities or actions in the real world. Its answer, impressive as it is, is derived purely from correlations among the words. Words have meaning only to a person who experiences life. We hold words as concepts derived from existence; ChatGPT holds words as correlations. Man's neural network is programmed by himself or herself; ChatGPT's neural network is pretrained using mathematical algorithms and data from external sources.

Even a simple fish engages in more self-programming ability than does ChatGPT. In a remarkable experiment using zebrafish larvae, Scott Fraser and coworkers at the University of Southern California have observed zebrafish larvae rewiring their biological neural networks in real time.[3] Fraser and his team employed genetic engineering on zebrafish, modifying them to generate neurons featuring a fluorescent protein marker tethered to their synapses. This marker protein fluoresces when exposed to the faint laser beams. This approach allowed the researchers not only to discern the precise locations of individual synapses but also to gauge their strength— brighter fluorescence indicating stronger synaptic connections.

To instigate a memory, Fraser and his team conditioned zebrafish larvae to associate a specific light with an uncomfortable increase in temperature. The zebrafish larvae acquired the behavior of

---

2 Asked about the origin of this hammer definition, ChatGPT-3.5 relied, "The response I provided is a generated output and not directly copied from any specific source."

3 William P. Dempsey et al., "Regional Synapse Gain and Loss Accompany Memory Formation in Larval Zebrafish," *Proceedings of the National Academy of Sciences* 119, no. 3 (January 2022), https://doi.org/10.1073/pnas.2107661119.

attempting to swim away whenever they encountered the associated light. The researchers imaged the zebrafish both before and after the fish acquired this learned behavior, examining changes in synapse strength and location. In those fish that did learn, synapses were selectively pruned from certain areas and strengthened in others. See Figure 1.

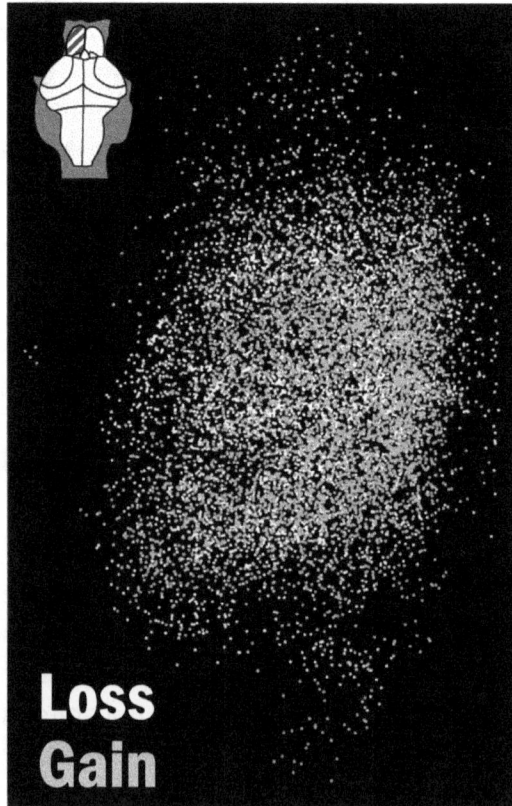

Figure 1. Map of lost (white) or gained (grey) synapses in the pallium region of a zebrafish larvae following tail-flick conditioning for fish number 1 in the dorsal view.[4]

---

4   Dempsey, "Regional Synapse Gain and Loss Accompany Memory Formation in Larval Zebrafish."

The brain's neural network does not have a designer. Our neural wiring is dynamic, internal, and individual.

Forty thousand synapse connections are made during every second during the third trimester of pregnancy. The complexity of the neural cortex increases following birth. The number of connections a neuron has peaks around the first year of life and gets reduced by a third thereafter. An adult has fewer neurons than a child; half the neurons created in the womb and during infancy die. To survive, a synapse or a neuron must be used. Unused neurons are pruned. The body cannot afford the dead weight of dormant neurons.

The first neural circuits to be programmed in a newborn baby are the visual cortex, to allow a baby to see, the motor cortex, to allow a baby to move, and the memory cortex, to allow a baby to learn. Concept formation comes with the first word, typically between the first and second year of life. From an early age, a child's character and view of life depends on what the child thinks about: *Do I love to play alone or with my sisters and brothers? Do I view Mommy and Daddy as kind or harsh? Is the world I see predictable or is it scary?* The perceptions formed in this early, implicit stage of thought are hardwired into the brain's neural circuitry.

Self-programming proceeds apace as the child grows. Thinking is hard and requires effort. *Do I want to focus on solving that math problem, or do I want to play a video game? Should I take that candy bar from Sally, or do I value her friendship more? Should I fess up and tell my dad that I broke the lamp, or do I blame it on Tommy?* Every thought a person has runs through a neural network and strengthens or lessens the connections. Free will is a consequence of biology: "Neurons that fire together come to be wired together." The connections in a brain are fortified when stimulated; connections atrophy when unused. A person who thinks about the reasons and consequences of his or her actions will develop very different brain circuitry from a person who evades the effort.

In a remarkable statement, Ayn Rand anticipated the physical basis of free will:

> That which you call your soul or spirit is your consciousness, and that which you call "free will" is your mind's freedom to think or not, the only will you have, your only freedom, the choice that controls all the choices you make and determines your life and your character.[5]

Thoughts control where we look, where we walk, and how we talk. No one doubts that Cathy Hutchinson (whom we introduced in *Question 31—How does my brain work?*) moved a robotic arm to pick up the bottle of coffee by thoughts—thoughts that generated signals picked up by electrodes implanted in her brain. Yet, the idea that our thoughts drive the direction of our thoughts is controversial. Determinists are loath to think that they are responsible for their own lives.

Free will is man's ability to program himself.

You control what you think about, and what you think about determines who you are. The knowledge "neurons that fire together come to be wired together" implies that what you think about matters. Where you focus your thoughts affects the wiring in your brain. Your attitudes, your actions, and your character are forged by what you think. The more you focus your mind on an idea, the more you strengthen the neurons associated with that idea in your brain. Free will is your ability to wire your brain as you like by focusing your thoughts.

---

5   FNI, 127.

A central aspect of self-programming is the self. Consciousness is awareness of existence. Free will is the ability to program oneself, and the self is awareness of the self-programmed brain. These three—consciousness, free will, and the self—are the essence of being human. No one has yet created an android that is conscious, that programs itself, or is aware of itself.

Our brains consist of neurons connected via synapses. These synapses fire or do not fire, analogous to the 1s and 0s in a digital computer. What is the magic sauce that converts the 1s and 0s in a biological brain into consciousness? Will we ever create a self-aware Dolores? A self-programmed Dolores? A Dolores with a will of her own?

The answer is likely yes, but the mechanism required to create a non-biological, yet conscious, Dolores remains inscrutable.

THE ANSWER

Can artificial intelligence make a robot conscious? Not at present. No one understands how the firing or the not firing of the neurons in our brains results in volition, in consciousness, and in a sense of self. No one knows how to make a robot program itself from scratch. Maybe someday someone will create a self-programming robot, a conscious robot, a robot with volition and a sense of self, but that day is not yet here.

IS IT POSSIBLE TO SPLIT THE BODY FROM THE MIND?

FREE WILL IN ACTION

Free will in thought is one thing, but what about free will in movement? Some people argue that Rand's definition of free will is purely a mental thing, that the body is not involved. After all, they say, Rand wrote in *Atlas Shrugged*:

"Free will" is your mind's freedom to think or not, the only will you have, your only freedom, the choice that controls all the choices you make and determines your life and your character.[1]

Is free will purely conceptual?

---

1   FNI, 127.

Galt's speech continues a few pages later:

> A body without a soul is a corpse, a soul without a body is a ghost...
> You are an indivisible entity of matter and consciousness.[2]

Rand was adamant: Man is an *integrated* being with both a body and a soul: a body composed of atoms; a soul produced by interactions among the atoms. Rand rejects:

> ...the soul-body dichotomy...mind versus heart, thought versus action, reality versus desire, the practical versus the moral.[3]

Free will in thought cannot be divorced from free will in action.

You have a thought; your body and eyes follow. It is impossible to read this sentence without focusing your mind on the concepts being presented *and your eyes focused on the words on the page.* Science does not yet know how thoughts originate from atoms, but we know how our thoughts make the atoms in our muscles move. Motion in muscles is directed by the action potentials coursing through our brains.

When Gustav Fritsch and Eduard Hitzig stimulated the cortex of a dog in 1870, they were astonished. The cortex—from the Latin *rind*—was thought to be the outer, wrinkled, protective layer of the brain, not central to brain function. Yet, when they probed the dog cortex with an electrode, they discovered each location in the motor cortex would cause twitches or spasms of small groups of muscles on the opposite side of the body. They had found the motor cortex, the region of the brain responsible for motion.[4]

---

2   FNI, 138, 142.
3   FNI, 51.
4   G. Fritsch and E. Hitzig, "Electric Excitability of the Cerebrum (Über die...

Subsequent researchers mapped the human cortex with various degrees of accuracy. By probing epileptic patients undergoing brain surgery, Penfield produced a detailed map of motor function in 1937.[5] The human motor cortex is a strip of neurons at the back of the frontal lobe. The muscular responses of the various regions of the motor cortex are shown in Figure 1.

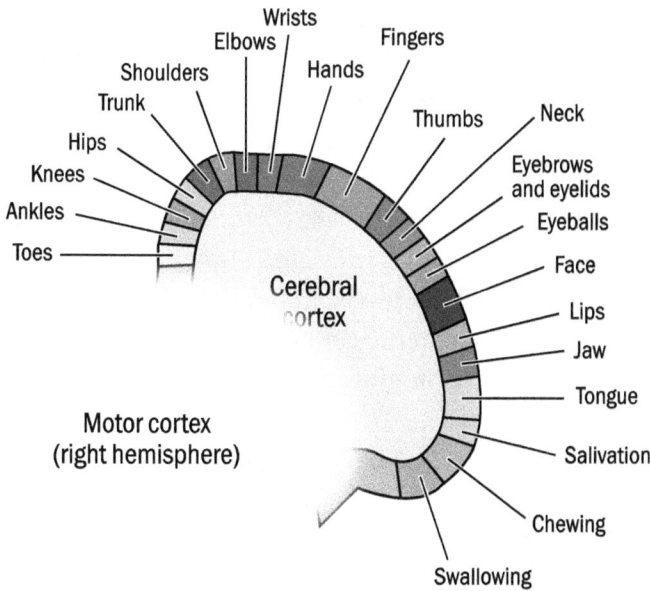

Figure 1. The locations in the motor cortex are responsible for motion.[6]

...elektrische Erregbarkeit des Grosshirns)," *International Classics in Epilepsy and Behavior: 1870* 15, no. 2 (June 2009): 123–30, 10.1016/j.yebeh.2009.03.001, reprint NIH-92-85, *Archiv für Anatomie, Physiologie und Wissenschaftliche Medicin* 37 (1870): 300–332, trans. Ted Crump and Shari Lama.

5   Wilder Penfield and Edwin Boldrey, "Somatic Motor and Sensory Representation in the Cerebral Cortex of Man as Studied by Electrical Stimulation," *Brain* 60, no. 4 (December 1937): 389–443, https://doi.org/10.1093/brain/60.4.389.

6   Image by CNX OpenStax (Wikimedia Commons, CC BY 4.0).

Penfield's brain map is useful—up to a point. It provides the "notes" of muscle action, but movement is a symphony. When you learn to swing at a tennis ball, your brain signals to combinations of muscles across your body, and it sends these signals in a coordinated fashion. Dozens of muscles are involved, such as your biceps and triceps, the anterior pectoralis (a band that crosses from the center of the chest to the arm), the anterior deltoid (a strip just in front of the armpit), and the broadest muscle in the back, the latissimus dorsi, which stretches from the lower back up to the underarm. The wrist and fingers are involved, as well as muscles in the legs. Natural motion is seldom, if ever, caused by the twitch of a single muscle.

To train the neural circuitry of the motor cortex requires persistence. An infant learning to walk will stumble and fall hundreds of times before coordinating the signals from the motor cortex to the muscles in the body correctly and in the right sequence. We are born tabula rasa not just in the contents of our ideas, but also in the actions in our limbs. We program ourselves from the bottom up.

Animals, too, employ self-programming to perform complex tasks.

Wolf cubs, for example, play at stalking prey to gain coordination for future kills. However, not all movement is programmed at the perceptual level. You could never teach a monkey to play tennis. Tennis has rules, rules require concepts, and only man exhibits concept formation. A child playing tennis for the first time may focus on swinging the racket correctly to hit the ball, but he or she is told the rules of the game. The return shot must clear the net and land in a certain rectangle to be successful; the ball should be hit inside the opposite rectangle so your opponent cannot return the ball.

The child absorbs these concretes and forms the concept "playing tennis." Through years of practice, swinging the racket becomes automatic, *but the rules of the game stay in the player's mind.* And a player has to think, "Should I hit this shot soft or hard, left or right, to win this point? Can I maneuver my opponent out of position so I can hit a passing shot?" Of course, this is done in an instant, without words, according to the concept of playing tennis the tennis player learned as a child.

Motion is the embodiment of free will. Watching a great tennis match is thrilling not only because of the physical accomplishment of the players but also because we can see the decisions players make in real time. Which player has the better strategy? Which player has the mental toughness to stay focused for the duration of the match? Which player will lose their mental focus and give up?

THE ANSWER

Is it possible to split the body from the mind? No. Man is an integrated being of body and soul. Free will is evident in what we think and what we do.

HOW IS CHARACTER BUILT?

THINKING CONSEQUENCES

Neural connections are modified by activity, as observed by the study of zebrafish at the University of Southern California and discussed in the response to the previous question. In humans, the subconscious brain does its thing, but the primary role of the conscious part of the brain is to think. The neural connections in the conscious part of the brain are modified by what we think about. However, thinking is not automatic. You must focus your mind on something to form a concept.

For man, the fundamental choice is to think or not:

Man's consciousness shares with animals the first two stages of its development: sensations and perceptions; but it is the third state, *conceptions*, that makes him man. Sensations are integrated into perceptions automatically, by the brain of a man or of an animal. But to integrate perceptions into conceptions by a process of abstraction, is a feat that man alone has the power to perform— and he has to perform it *by choice*. The process of abstraction and of concept-formation is a process of reason, of *thought*; it is not automatic, nor instinctive, nor involuntary, nor infallible. Man has to initiate it, sustain it, and bear responsibility for its results. The pre-conceptual level of consciousness is nonvolitional; volition begins with the first syllogism. Man has the choice to think or to evade—to maintain a state of full awareness or to drift from moment to moment, in a semi-conscious daze, at the mercy of whatever associational whims the unfocused mechanism of his consciousness produces.[1]

The phrase "in a semiconscious daze" is not accurate. A person can be conscious or unconscious; "semiconsciousness" occurs only when we are sleepy or drunk. The problem isn't focus; it is focusing on the wrong things. One has "the choice to think or to evade," but the "evasion" is a consequence of thinking the wrong things. Incorrect thinking and incorrect character building are caused by rationalization, not considering the full consequences of one's thoughts. Forming a negative character trait results from too narrow of a focus on a bad idea.

A murderer isn't avoiding the process of thinking. He is focused on evil thoughts. His evasion is not considering the broader picture: Taking a human life is evil. An honest man—a non-murderer—may experience the thought, "If I kill someone and steal his money, I'll be rich," but he dismisses it immediately. He considers the broader context of the act. His self-programming to commit murder is nonexistent.

---

1   FNI, 14.

THE ANSWER

How is character built? Your character is a consequence of where you focus your mind. Thinking about something modifies the neural connections in the brain, reinforces your thought patterns, and locks in good or bad behavior. Each person is responsible for what they think about and, hence, for the type of person they become. We are born tabula rasa and program ourselves from the ground up. The story of your life has an author—it is you!

WHAT IS THE PURPOSE OF ART?

WHAT ART IS

L
ast night I watched a romantic comedy movie about the misunderstandings, confused motivations, and misplaced attractions of two potential lovers. I enjoyed it thoroughly and laughed at the comic situations. This morning, I woke up feeling optimistic and happy.

Why do we watch movies, read novels, look at paintings, listen to music, and play computer games? Art seems to have no practical value. Yet great art captivates us, while bad art makes us feel we are wasting our time.

Ayn Rand said our need for art derives from our conceptual consciousness:

The source of art lies in the fact that man's cognitive faculty is conceptual—i.e., that man acquires knowledge and guides his actions, not by means of single, isolated percepts, but by means of abstractions.[1]

We discussed conceptual abstractions in answering *Question 3—What are concepts?* and learned that a word is a label for a concept that represents an unlimited set of concretes of a certain kind. When we speak, we form a conceptual chain involving many levels of abstraction, all the way from sense data—that curveball was fast—to intellectual knowledge—how much did the air pressure drop on the lee side of the ball? While conceptual concepts are complex, they are derived from things that exist. Normative concepts are even harder to grasp. These concepts designate some actions as good, desirable, or permissible, and others as bad, undesirable, or impermissible.

Normative concepts concern alternatives about how to behave.

Normative concepts do not arise in a vacuum. They depend on earlier concepts: What exists and how do you know it? Ayn Rand explains:

Is the universe intelligible to man, or unintelligible and unknowable? Can man find happiness on earth, or is he doomed to frustration and despair? Does man have the power of choice, the power to choose his goals and to achieve them, the power to direct the course of his life—or is he the helpless plaything of forces beyond his control, which determine his fate? Is man, by nature, to be valued as good, or to be despised as evil? These are metaphysical questions, but the answers to them determine the kind of ethics men will accept and practice; the answers are

---

1   RM, 10.

the link between metaphysics and ethics. And although metaphysics as such is not a normative science, the answers to this category of questions assume, in man's mind, the function of metaphysical value-judgments, since they form the foundation of all of his moral values.[2]

Ethics—the way a person treats other people—thus rests on an individual's thoughts about metaphysics. He or she may not realize it, but there is no escape from reality.

This is where art enters. Someone may say that they watch comedy to escape reality. But no, all comedy is metaphysical, as is all art. We laugh because the characters in the comedy are confused about the world as it exists.

Watching a comedy is not like everyday life. The author of the comedy has rearranged happenings we are familiar with into a new and unexpected form. This recreation of reality is what captivates us. We are captivated by recreations of reality, from comedy and drama to music and painting, provided the art is high quality.

Ayn Rand defined art as:

> Art is a selective re-creation of reality according to
> an artist's metaphysical value-judgments.[3]

According to this definition, journalism is not art. Journalism, properly done, reports happenings as they occurred in real life. A novel, however, is art because the author has created an imaginary world emphasizing how he or she thinks about life.

---

2   RM, 12.
3   RM, 13.

WHAT ART DOES

Art supports man's conceptual level of awareness in three essential ways. The first exemplifies the consequences of behavior that is hard to absorb from everyday life.

Art makes normative concepts real.

Ayn Rand observed:

> Art is a concretization of metaphysics. Art brings man's concepts to the perceptual level of his consciousness and allows him to grasp them directly, as if they were percepts.[4]

Second, art paints life as it should be. Good art, the kind that makes our hearts sing, carries us along with a sense of purpose and hope. This fuels our ambition, orients us toward right actions and away from committing wrongs, and helps frame our sense of life:[5]

> Art is the indispensable medium for the communication of a moral ideal.[6]

Objectivists derive inspiration from the ideal men and women in *The Fountainhead* or *Atlas Shrugged*, while others find these characters threatening:

---

4   RM, 13.
5   "A sense of life is a pre-conceptual equivalent of metaphysics, an emotional, subconsciously integrated appraisal of man and of existence." RM, 17.
6   RM, 15.

Art is man's metaphysical mirror; what a rational man seeks to see in that mirror is a salute; what an irrational man seeks to see is a justification...[7]

Third, art is a mechanism for training your emotional neural network. Every work of art you contemplate is fed into your subconscious as you sleep. A wise person chooses his art carefully. Positive life-affirming books and films enhance your happiness; soul-destroying stories leave you in despair.

THE ANSWER

What is the purpose of art? The purpose of art is threefold: to make normative concepts real, to help shape a person's sense of life, and to provide individuals with emotional training materials to program his or her subconscious.

---

7 RM, 30.

HOW DO WE KNOW RIGHT FROM WRONG?

MORALITY IS A CONSEQUENCE OF EXISTENCE

Programming ourselves creates a problem. We must direct the course of our lives, but have no innate guide. Our every thought is error prone. For thousands of years, people believed the earth was flat at the center of the universe, and the stars revolved around the earth. Yet the fact that the earth is round, not at the center of the universe, and rotates is easily checked. Morality— the guide to the values each person uses to make decisions about the choices in their lives—is not so easily seen. To sort through the abstract issues involved in establishing an existence-based moral code, we adapt Rand's question, "What, in reality, gives rise to the need for a moral code?"

The first part of the answer is a consequence of the conjectural nature of concepts. Without an innate code, man must determine these values by thinking. And thinking is far from automatic. Just as we teach our children that the earth is round and spins on its axis, we must teach our children what is right and what is wrong. The moral guide we provide should be in accordance with reality, but moral concepts we teach are many steps away from sense data. We need to find the logical chain, the chain that traces moral concepts all the way back to observables.

With respect to morality, the first observable is the difference between living and nonliving entities:

> There is only one fundamental alternative in the universe: existence or non-existence—and it pertains to a single class of entities: to living organisms. The existence of inanimate matter is unconditional, the existence of life is not: it depends on a specific course of action. Matter is indestructible, it changes its forms, but it cannot cease to exist. It is only a living organism that faces a constant alternative: the issue of life or death. Life is a process of self-sustaining and self-generated action. If an organism fails in that action, it dies; its chemical elements remain, but its life goes out of existence. It is only the concept of "Life" that makes the concept of "Value" possible. It is only to a living entity that things can be good or evil.[1]

> Good and evil, right and wrong. To a living organism, the right course of action promotes its life; the wrong course of action harms its life.

An inanimate object can have no goals. It simply is. Living organisms are different:

---

1   FNI, 121.

Only a living entity can have goals or can originate them. And it is only a living organism that has the capacity for self-generated, goal-directed action. On the physical level, the functions of all living organisms, from the simplest to the most complex—from the nutritive function in the single cell of an amoeba to the blood circulation in the body of a man—are actions generated by the organism itself and directed to a single goal: the maintenance of the organism's *life*...if an organism fails in the basic functions required by its nature—if an amoeba's protoplasm stops assimilating food, or if a man's heart stops beating—the organism dies. In a fundamental sense, stillness is the antithesis of life. Life can be kept in existence only by a constant process of self-sustaining action. The goal of that action, the ultimate value which, to be kept, must be gained through its every moment, is the organism's *life*.[2]

Tying *value* to *life* establishes the first link in the chain between observables and morality. Acts that preserve its life are good for the organism; acts that destroy its life are evil. Life thus becomes the standard of value for an organism, the standard by which all lesser values are evaluated. Life is an end in itself.

> Your life is the standard of value by which right and
> wrong are evaluated. That which furthers your life
> is good, that which harms your life is evil.[3]

Setting your life as the standard of value is easy. The hard part is figuring out *what* is in your self-interest. Is it in your self-interest to kill other people? Not if you value your life. Life expectancy is short

2   VOS, 4.
3   See VOS, 13 and AS 1014.

in a society where murder and violence are rampant. Living in a peaceful society, a society where murder and mayhem are outlawed, prolongs your life—to say nothing of increased harmony and happiness. The moral principle "Thou shall not kill" is in your rational self-interest.

The commandment "Thou shall not kill" provides an example of moral teaching, but it is limited in scope. Broadening morality to cover all of life's significant decisions requires a focus on man. Man's life is his standard of value, but what is man's essential means of survival?

Every species has an evolutionary advantage that allows it to survive. The tiger has its ferocious bite, the deer its swiftness, the turtle its shell, and the skunk its smell. Man has his mind. Our evolutionary advantage is the ability to form concepts. Man is weak in tooth and claw, slow afoot, and soft to eat. He would not last long on the planet if he did not think.

Survival *qua man* means survival of man's ability to think. It does not mean survival in the body alone. A man may survive in a coma or in a gulag fighting for his life, but, in these states, he is not able or free to think and act on his thoughts. Further, man must think long term—the decisions he makes today will affect his survival for the rest of his life. Rand writes:

> "Man's survival *qua man*" means the terms, methods, conditions and goals required for the survival of a rational being through the whole of his lifespan—in all those aspects of existence which are open to his choice.[4]

This idea may be expressed in logic as:

---

4   VOS, 14.

Major premise: Man's life is the standard of value.
Minor premise: Reason is essential to man's life.
Conclusion: Reason is essential to the standard of value.

The moral is thus that which promotes man's ability to reason, the immoral that which destroys it.

RIGHT AND WRONG

The freedom to think is the essence of morality, but how does this translate into acts? Why are some acts right, while other acts are wrong?

Thinking is an individual activity. No one can think for you. Your ability to think can be impeded by self-destructive acts, such as drinking, drugs, and reckless behavior, or by force used by others who say, "Do it my way or else!"

The first category of moral imperatives is to respect yourself. Your mind is your treasure. Do not destroy it. Every conscientious parent tells their child, "Don't drink too much, don't take drugs, and think before you act." These moral commands are derived from man's need to preserve his mind.

The second category of moral imperatives is the need to protect you and your thoughts from the brutality of others. Men have tried to subjugate others since the dawn of time. Murderers, thieves, kings, and dictators use thuggery to get their way. While a political prisoner might dwell on the injustice of his circumstances, his ability to act and change them is taken away.

Physical force is the distinguishing characteristic of this second category of moral imperatives. Not force applied to retaliate against a murderer or a thief, but the initial threat from the perpetrator. Immorality derives from the *initiation* of the use of force:

The basic political principle of the Objectivist ethics is: no man may initiate the use of physical force against others. No man—or group or society or government—has the right to assume the role of a criminal and initiate the use of physical compulsion against any man. Men have the right to use physical force only in retaliation and only against those who initiate its use. The ethical principle involved is simple and clear-cut: it is the difference between murder and self-defense. A holdup man seeks to gain a value, wealth, by killing his victim; the victim does not grow richer by killing a holdup man. The principle is: no man may obtain any values from others by resorting to physical force.[5]

The moral commandment "Don't initiate the use of physical force against others" is manifested by what we teach our children in myriad ways: "Johnny, don't take Billy's toys," "Tom! Don't fight with Sam!," "Suzi, let Becky play with the group," and a thousand more. In each case, the right thing to do is to let the individual mind be free.

THE ANSWER

How do we know right from wrong? Morality derives from man's need to think and act on his thoughts. It is right to allow the mind, and the consequences of the mind, to flourish. It is wrong to destroy the mind through drugs and drink, or to initiate the use of force to prevent someone from acting on his noncoercive thoughts.

---

5   VOS, 24.

IS THAT EVERYTHING?

THE ULTIMATE QUESTION

With this question we have reached our limit of 42 Ultimate Questions. The questions included in this book are *Ultimate* in the sense they concern life and death, existence and nonexistence, and knowledge versus superstition. The most important question on this list is whether existence or consciousness is the primary fact in the universe. All other questions and answers depend on this choice.

The objectivist takes the view that existence exists independently of consciousness, and that consciousness is awareness of existence, not its creation. This leads to metaphysics (what is existence?) and to epistemology (how do we know it?). In most areas, metaphysics and epistemology are intertwined. Beyond accepting the basic proposition that *existence exists independently from consciousness*, metaphysics equates to science. *What exists?* is a scientific question, but this question cannot be answered without a proper epistemology.

Questions such as *Question 19—Do we live in a clockwork universe?* and *Question 22—Do we live in a multiverse?*, are fundamentally epistemological questions even though their focus is on physics.

> Man will never know everything, but with the
> proper guide, he can know more and more.

While the forty-two questions and answers in this book cover a broad swath of knowledge, they do not answer everything. Knowledge is open-ended and ever growing. This guide is objectivism, the philosophy developed by Ayn Rand.

THE ANSWER

Is that everything? The forty-two questions and answers in this book address the broadest issues man faces: what exists, how he knows it, and how he should behave. Many subjects are examined in this book, but many others are left out. No book can cover everything, but hopefully, life's most important Ultimate Questions have been asked and answered.

ACKNOWLEDGMENTS

Writing this book has been a journey. From its beginnings in attempting to explain some scientific concepts to objectivists, to assembling a quantity of diverse materials under the working title *Objective Science*, to the complete work you hold in your hands, I was prodded, shaped, and informed by a great many people. Family and friends were the first to contribute, followed by professional acquaintances, and finally my two editors.

Of the many family members and friends who read early copies of the manuscript and provided feedback, I would like to extend special thanks to my son Patrick, daughters Linda and Yvette, brother Nick, nephew Chris, son-in-law Ty Hosler, and friends Rick Edson and Al Stiner for helpful and incisive comments. Zoltan Oltvai came into the process much later, but also helped to steer the book in the right direction. Their feedback and encouragement helped me to keep the project going.

Special appreciation is due Lee Pierson, who read an early draft of this book and made hundreds of suggestions for improvements. His insight into the nature of consciousness made significant improvements to the book. Keith Lockitch of the Ayn Rand Institute provided comments that helped improve the structure of the book. Stephen Hicks of the Atlas Society helped fine-tune some of the history and philosophy discussion. David Kelley, also of the Atlas Society, helped me to work through some issues with space and time, as well as prompted me to take another look at the problem of induction. Thank you, Lee, Keith, Stephen, and David.

Finally, I want to express my thanks to my editors, Mark Chait and Tashan Mehta of Scribe Media. They both invested considerable thought into this project. Their work helped me chart a new course across the sea of knowledge on which I embarked. They calmed the seas of my turbulent original draft. Thank you, Mark and Tashan, for helping me along my journey.

ABOUT THE AUTHOR

Dr. Zoltan Cendes is professor emeritus in the Electrical and Computer Engineering Department at Carnegie Mellon University in Pittsburgh, Pennsylvania. He is a fellow of the Institute of Electrical and Electronics Engineers and was elected to the National Academy of Engineering in 2021 *"for contributions to theory, development, and commercialization of electromagnetics simulation software."* Cendes is founder of Ansoft Corporation, a trailblazer in engineering simulation software. Under his leadership, Ansoft's innovative use of the finite element method revolutionized electrical product design and attracted major clients that include Intel and Apple. Following a successful IPO, Ansoft was acquired by ANSYS for $900 million, enhancing its global impact in engineering simulation. A pioneer in engineering and mathematics, Zoltan inspires the business and scientific worlds with his deep insight into the intersection of technology, philosophy, and reality.

www.ingramcontent.com/pod-product-compliance
Lightning Source LLC
Chambersburg PA
CBHW051709020426

42333CB00014B/911